Praise for *The Ocean in a Drop*

'The earth is ailing. What is the diagnosis, and where is the cure? Roz Savage is dauntless in her quest for answers. Strap in and join her for a riveting ride on *The Ocean in a Drop*. A world-class athlete, she rows the Pacific alone, with an unblinking eye on the symptoms of the patient. An Enlightenment intellectual, she pursues a diagnosis based on reason, and on sciences varying from politics and economics to cognitive neuroscience and evolutionary biology, all the while knowing that if reason is consistent then it must be incomplete. A compassionate human, she inspires us to embrace our innate intelligence that transcends the limits of reason, and follow its guidance to unexpected remedies. If we heed her prescription, our prognosis is bright.'

Donald Hoffman, author of *The Case Against Reality*

'This is a book by, of and for this moment. A paean to human resilience and courage, it shows what happens when we cast off the shackles of our normative behaviours and discover who we really are, attend to our inner voices of integrity and authenticity and stretch ourselves to the edges of our being ... Roz Savage paves the way towards a different future with the blistering light of deep personal exploration, and the authentic voices of some of our planet's deepest thinkers. This book is essential reading for anyone who wants to see the more beautiful world our hearts know is possible come into being.'

Manda Scott, author of the *Boudica: Dreaming* series

'This well-researched and readable book may well give encouragement to readers for a change in our conscious lifestyle. The author points out that our civilisation is sick. To effect a cure, it is helpful to examine the symptoms and underlying causes. The good news is that we are nonetheless fundamentally healthy. Roz Savage sets out not only to lucidly present the ills that beset our world – environmental, economic, political, social and so on, but she uncovers the psychological, neurological and spiritual causes of our problems. Based on this analysis with characteristic intrepidity she presents her vision for a future redemption when humanity will finally turn the corner and regain its healthy values of basic sanity. Despite humanity's pathetic track record to date, Ms Savage dares to nurture sanguine expectations of our ultimate reform.'

Jetsunma Tenzin Palmo, Buddhist nun

'Anyone who rows solo across three oceans deserves our undiluted attention. Beyond that, this magical book distils Roz Savage's hard-won wisdom at a time when we all must learn to see ourselves and our worlds from outside. *The Ocean in a Drop* alternates between the alarming, the challenging and the uplifting. Formidable forces are now at work, for better or worse, and, as the book concludes, "miracles might just happen".'

John Elkington, founder and chief pollinator at Volans, and author of *Green Swans*

'*The Ocean in a Drop* vividly shares a wonderfully authentic and transformational journey, of both inner and outer discovery that is radically relevant for all of us for how we see ourselves and the world at this critical moment of choice. While its clear-eyed and yet compassionate appraisal of how our current systems and behaviours, based on separation and dominance, have brought us to an existential crisis, its greatest gift is to reveal and empower the potency and potential, through our conscious evolution and a new and unitive narrative, of how our existential emergency can be transformed to the emergence of a world that works for the good of the whole.'

Dr Jude Currivan, cosmologist, author of *The Cosmic Hologram* and *The Story of Gaia* and co-founder of Whole World-View

'Great wisdom can be found in the convergence of science and spirituality. Roz takes this wisdom and applies it to our seemingly intractable global challenges. She paints a picture of how we can use our attention, intention, and innate gifts to achieve an important paradigm shift that would propel us towards a beautiful vision of a thriving future.'

Chade-Meng Tan, author of *Search Inside Yourself* and *Joy on Demand*

'A thoroughly researched document that gives the reader a clearer understanding of why the world is as it is – the economics, social, psychological and cultural forces at play. Roz presents a vision of how the future can be better while inspiring and encouraging us that we have a purpose, a role to play and the power to make a positive difference. This book is well timed and needed, a life preserver in our chaotic times.'

Will Steger, educator and explorer

The Ocean in a Drop

Navigating from Crisis to Consciousness

ROSALIND SAVAGE

First published 2022

FLINT is an imprint of The History Press
97 St George's Place, Cheltenham,
Gloucestershire, GL50 3QB
www.flintbooks.co.uk

British Library Cataloguing in Publication Data.
A catalogue record for this book is available from the British Library.

ISBN 978 0 7509 9969 4

Typesetting and origination by The History Press
Printed and bound in Great Britain by TJ Books Limited, Padstow, Cornwall.

Trees for Life

Contents

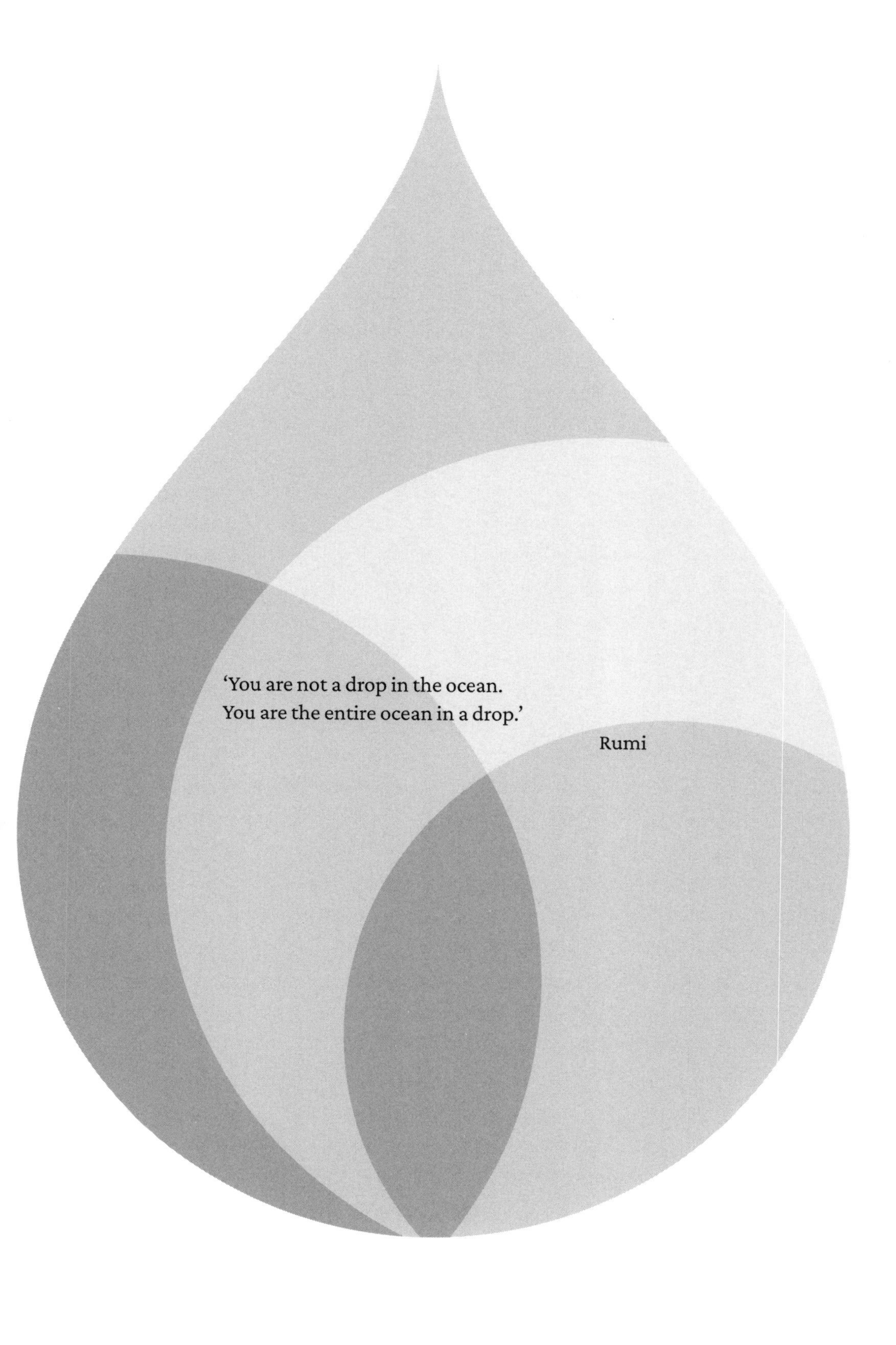
'You are not a drop in the ocean.
You are the entire ocean in a drop.'
Rumi

Prologue

I was awoken by my bed suddenly lurching several feet to the side and tilting sixty degrees. I groaned as I returned from a pleasant dream about a buffet table creaking under the weight of all my favourite foods and remembered where I was: alone on a 23-foot rowboat in the middle of the Atlantic Ocean. Deeply resenting the wave that had smashed into the side of my boat and wrenched me away from the smorgasbord of delights, I rolled over on my narrow bunk and closed my eyes, trying to return to blissful unconsciousness. But reality had smashed in as brutally as the wave, and sleep was gone.

I sighed and turned on the GPS to see where I was. During the hour I'd been asleep, the winds and currents had carried me about a mile towards my destination. One mile out of a total of 3,000, from one side of the Atlantic to the other. Not bad, but if I'd been rowing instead of dozing I would have covered two or maybe even three miles. But the distance ahead seemed so vast, I could make up for this wasted hour later ... Couldn't I?

I paused. An uncomfortable truth was niggling at the edge of my awareness. I gave it a moment to reveal itself.

This wasn't really about the difference between one mile, or two or three. This was about who I was and how I wanted to show up for this enormous challenge I'd set myself. Did I want to be the kind of person who slacked off and skipped rowing shifts when I felt tired, overwhelmed or in pain? Or did I want to be the kind of person who committed fully, did my best and stuck to my rowing schedule with discipline and determination? With every choice I made – whether to row or not to row – I wasn't just crossing an ocean. I was creating a story about myself. I would become the sum total of the seemingly tiny choices I made every day, each one inconsequential in itself, but adding up to a story of what it means to be me.

Who am I?

Who do I want to become?

How do I show up when nobody's looking?

I knew what story I wanted. It wasn't yet true, but this was in my hands. I could make it come true by making different choices – choices that might seem harder at the time, but would bring ease and peace at the end of each day when I could log the miles covered and know I'd done all I could to get closer to my vision – a vision not just of reaching the other side of the ocean, but of crafting a new identity for myself, no longer a burned-out management consultant, but an adventurer and explorer of the oceans, of the world.

I shook myself out of my reverie, pulled on my rowing gloves and sunhat, grabbed my seat pad, and headed out of the cabin to the rowing seat to continue creating my new story, one oarstroke at a time.

As we approach the quarter-way point of the twenty-first century, humanity is being challenged to create a new story about who we are. When we contemplate the myriad wrongs of the world – such as inequality, racism, social injustice, climate change, plastic pollution, ocean acidification, habitat destruction, poverty, conflict, corruption, authoritarianism, polarisation and discrimination – we might feel despair and desperation. Many of us have bought into a story of separation, isolation and existential fear that paradoxically creates the very situations we're trying to avoid. While we choose to believe that each of us is just an inconsequential drop in the ocean, we do a great disservice to ourselves, our species and our planet. It might be a comforting story, because identifying as a powerless droplet gets us off the hook of doing anything about global challenges that can, admittedly, seem overwhelming. But it's not a true story, and it's not an empowering one.

In the midst of ongoing turmoil, and having weathered many storms of grief and despondency, I have found hope. But as the environmental entrepreneur Paul Hawken said on my podcast, hope can be the pretty mask of fear, so we need the right kind of hope: a trenchant tenacity and beautiful audacity; a surrender to *what is*, while also holding fast to a vision of *what can be*; and the deep knowing that we have the power, energy and agency to make a difference through the choices we make. It's this kind of hope I offer and that this book is about: a hope grounded in a new story about what it means to be human at this critical juncture in our species' history; a story that gives us back our courage, dignity and power; a story that can lead us into a new paradigm of unity, connection and love.

This book is about the course I've navigated over the last twenty years to try and find that better story – my questions, frustrations, inspirations and occasional quantum leaps in understanding. Like many adventurers, I began by studying the paths forged by others – those who have explored questions around the environment, economics and the ways in which humans have chosen to live, individually and collectively. I've combined and recombined and riffed on these concepts to arrive at a plausible explanation of where we are now and to create a vision of a possible future. This book is the account of my personal voyage of discovery, my adventures and misadventures in envisioning a better world.

As you will see, this book is ambitious in its scope and bold in its imagination. I throw a lot of ideological and metaphysical spaghetti at the wall, and some of it will stick for you, some of it almost certainly won't. To mix my metaphors, I just ask that you don't throw out the baby with the bathwater. Even if you violently disagree with some of the ideas presented here, please persevere with an open mind. I intend to start a conversation, not to have the final word. I cherry-pick from the books I read, choosing the juicy fruits, adopting and adapting them to my own needs, and discarding the rest. I trust you to do the same.

For sure, I'm not claiming to be the hero arriving with the eleventh-hour solution to all our problems. I'm not even saying we're going to make it through the multi-tentacled crises currently engulfing the world. It's up to all of us to forge our future together. Crisis is an opportunity for transformation. Good things are possible, and together we can make them probable. Together, we can create a clear vision of the future we want and hold that vision as our north star as we navigate the uncharted, turbulent waters of the future.

PART ONE

THE BAD NEWS

'The seven social evils are:
Politics without principle.
Wealth without work.
Pleasure without conscience.
Knowledge without character.
Commerce and industry without morality.
Science without humanity.
Worship without sacrifice.'

From a sermon given by Frederick Lewis Donaldson
in Westminster Abbey, 20th March 1925

Chapter 1

Environment

> 'There's nothing fundamentally wrong with people. Given a story to enact that puts them in accord with the world, they will live in accord with the world. But given a story to enact that puts them at odds with the world, as yours does, they will live at odds with the world. Given a story to enact in which they are the lords of the world, they will *act* like lords of the world. And, given a story to enact in which the world is a foe to be conquered, they will conquer it like a foe, and one day, inevitably, their foe will lie bleeding to death at their feet, as the world is now.'[1]
>
> Daniel Quinn

A hailstorm hit just as we left our campsite, high in the Andes above Cusco, the hailstones pinging painfully off ears, noses and hands. The first section of our trek was mostly uphill, a stiff climb, but the old woman leading our procession strode on undaunted. The pace was being set by someone twice my age and pride forced me to keep up.

As the hailstorm abated, we were left with a dramatic skyscape of dark clouds and bright sunshine, the sun's rays picking out distant mountains in a variety of colours and textures – red dust, green grass, white snow. My self-appointed guide, Abel, had promised me that this was a *camino bonito* – a much prettier walk than the dusty path that had led us to the sacred mountain of Ausangate several days before – and he was right. Our little troupe of pilgrims filed across the high *altiplano*, headed by a brother carrying the flag of the Inca, while we dutifully followed the fluttering rainbow.

When we reached a high pass, Abel pointed out a plume of smoke rising from a distant valley, signalling the village where we would rest for a few hours. The trail led straight to it, but to our left I could see a herd of alpacas and a flat plain of serpentine pools that beckoned to the photographer in me. I told the others I would catch up with them later and headed off alone to spend a pleasant hour taking pictures, watching the late afternoon sun raking across the landscape, backlighting the soft fur of the alpacas into golden halos. The solitude and peace of the mountains were a welcome respite from the noise and confusion of the last few days, when 30,000 pilgrims had crowded into a high hanging valley for the festival of Qoyllur Rit'i and had danced morning, noon and night – possibly out of practicality as much as piety, as it was icy cold up by the glacier and dancing was the best way to fend off hypothermia.

Feeling calm, restored and reconnected with nature, I strode purposefully down into the village, following a cascading stream through a deep gully, and rejoined the others. Spirits were high when I arrived at the pilgrims' camp. A couple of lively *ukukus* (men dressed as bears, the designated pranksters of the pilgrimage) had run up the mountainside, and everybody else was watching them, shouting encouragement and laughing at their antics as they slithered and tumbled their way back down the slope.

We had only about four hours to rest before we would be back on the trail again. Abel had explained to me how, before leaving Ausangate, he would go to a rock near the chapel where an angel and a devil – not real ones, I later established, but humans in costume – would tell him what time we were to start the final leg of our pilgrimage. Our appointed hour was ten o'clock that night.

I made myself comfortable in the lee of a drystone wall, setting out my camping mat and sleeping bag on top of a plastic poncho. I pulled tight the drawstring of my sleeping bag, leaving only a small air hole, and snuggled in, feeling snug and smug in my cosy cocoon, anticipating a restful sleep.

No sooner had I settled down than the band struck up at top volume 6 feet away from my head. A minute after that, there was a sudden and unexpected attack on my cocoon by a couple of unseen perpetrators, most likely *ukukus*, who ran by giving me a passing prod, possibly to mock my first-world-style comfort. But after that, I slept soundly for four refreshing hours.

I was woken by somebody kicking my feet, urging me to get ready. I emerged from my warm den to find a thick coating of frost on the outside of my sleeping bag. I silently cursed the angel of the rock for the untimely wake-up call, packed up hastily, and headed into the moonlit night.

It's two o'clock in the morning and at last I understand why I'm here. All I've seen and heard over the last few days finally clicked into place when we stopped for a rest on our nocturnal trek and I watched as a line of pilgrims passed by, silhouetted against the full moon, the campesina *women distinctive figures in their multilayered woollen skirts. For the first time, I understood that I am truly on a pilgrimage. Qoyllur Rit'i is a pilgrimage in itself, but so is my whole trip to Peru. My fellow pilgrims are here to test themselves physically and to recharge their spirituality, and so am I. I've been trying to make sense of this through words, which hasn't worked, especially as the pilgrims' first language is Quechua and mine is English, and none of us speak Spanish well. But my new understanding transcends words. Even though we come from different worlds, I finally know that I am part of this brotherhood of pilgrims; I belong here.*

This deep knowledge of our unity of purpose stayed with me as we marched on through the night, guided by the stars of the Southern Cross, the men at the front of our little procession bearing two embroidered standards on frames shaped like crucifixes. The bells on the costumes of the *ukukus* tinkled softly as they walked, and a distant drum beat the now-familiar one, two, three-and-a-four.

There was sacred purity in our nocturnal trek. I could see enough of our surroundings by the bright light of the moon to appreciate the Andean grandeur, but not enough to be distracted by it. The mood was contemplative, meditative, lulled by the rhythms of the drums and the bells. There was the occasional outburst of banter and laughter, but mostly we walked in silence. Where the path was rough and stumbly underfoot, or where the shadow cast by a hill blotted out the moonlight, I turned on my head torch to light my way, but I preferred to walk, as the others did, without artificial light. I felt calm and content, part of my mind focused on walking surefootedly, but the rest free to wander, free in the timelessness of an obscure hour in a strange land on a beautiful moonlit night.

It's half-past four in the morning and we've stopped for another rest break. My fellow pilgrims are snoring softly around me, but I'm not sleepy. I'm leaning back against my rucksack, smoking a cigarette to calm my hunger, contemplating the sky. Stars are glimmering through the gaps in the fish scales of cloud. It bodes well for a good sunrise, which would be a welcome reward after walking the best part of fifteen miles to see it.

The Catholic veneer of the first few days of the pilgrimage has melted away, and now I feel the ancient, animistic Inca faith emerging. We have appeased the mountain spirits who have the power to grant feast or famine, fortune or failure. We've paid homage to the various elements of earth and sky, and now we're walking by the light of the moon and stars to see the sun rise on the sacred mountain of Ausangate.

The festival at dawn was all the more dazzling for its contrast with the quiet calm of the night. In the shady morning twilight, shaggy coats, magnificent feathered headdresses and masks magically appeared out of makeshift rucksacks and were swiftly donned. The pilgrims lined up along the brow of a hill, their costumes glowing in jewel colours in the soft predawn light. Ranks of flags – the red and white of the Peruvian national flag and the rainbow stripes of the Inca flag – fluttered from handheld banner poles. We faced towards Ausangate, now in the far distance, waiting for the sun to rise. As the sun peeked over the horizon, it slanted across the mountain vista to strike the distant snowcap. This was the signal for the start of an exuberant pageant, the participants charging back and forth across the crest of the hill, shaking hands with their fellow pilgrims and greeting the sun. I wandered up and down the line, mesmerised, absorbing the impassioned energy of the celebrations.

According to Abel, the final hilltop celebration was about the sun first, and *El Señor* (the Catholic Christ) second – and a distant second, I would say. The pilgrims gather at dawn to give thanks to the sun for light and health and food in an energetic celebration of life, reciprocating the life-giving energy of the sun.

As the hilltop celebrations were drawing to a close, I walked alone down to the village while my compadres went to the chapel to complete their devotions. I looked back at where I had come from, at the celebrants streaming down the hillside in two intersecting diagonal lines, a huge human cross against the green grass, skipping and running with infectious exuberance and sheer wholehearted passion for life.

I was ambushed by an unexpected upwelling of emotion, an enormous affection for this magical land of Peru and its people. Words failed me, for once. I couldn't articulate, even to myself, why my heart felt so full. It could have been the unbridled ecstasy of the pilgrims, or the beauty of the starlit trek, or the groundedness of sleeping on the earth for the last few nights, or the sense that I was coming home to myself and my inborn love of adventure, or appreciation for the series of serendipities and synchronicities that had brought me to this country and this valley and this new life I was crafting for myself. All I knew for sure was that I was overflowing with bliss, and I walked down the hill with tears of joy in my eyes.

Peru may seem like a strange place to start this story, but it's as good a place as any. It was 2003 and I was 35 years old, recently jobless and divorced, when I spent three and a half months in Peru. It was the first time I'd been backpacking – mostly unscheduled, spontaneous and free – and the journey set many of the themes that have dominated my life ever since.

I can trace my deep and abiding concern for our Earth back to this trip, and specifically back to that glacier on the sacred mountain above Cusco. My fellow pilgrims told me that each year they had to trek a bit further to reach the glacier at the summit of Ausangate, the focus of their celebrations, because it was retreating as it melted. This was my first encounter with the real-world impacts of climate change, and it hit me hard. I felt I absolutely must do something, anything, to raise awareness and spark action, as a matter of the utmost urgency. It may seem like a major leap of logic to get from melting glaciers to ocean rowing, but I can only say that it was the least logic-driven decision I've ever made, and also the best. The day when the idea seized hold of me was about six months after I'd resolved to take action, six months of growing frustration as I sensed the clock ticking, every passing day bringing more and more environmental damage, yet completely at a loss as to what I could do to contribute to the cause. I had wracked my brains, but found no answer. Turns out, my brains are rarely the best source of truly original ideas. The best so-called brainwaves come from somewhere else.

Inspiration finally struck when I was on a long drive north to my parents' home in Yorkshire, thinking of nothing in particular, just autopiloting along with my brain in neutral. I was aware of the obscure and masochistic sport of ocean rowing via a friend, Dan Byles, who had rowed across the Atlantic with his mother in 1997 in the first ever Atlantic rowing race, but it hadn't struck me as a fun or sensible thing to do. Yes, I had two half-blues for rowing for Oxford in the annual races against Cambridge, but rowing across oceans was most definitely not on my bucket list. Frankly, I was terrified of the ocean, having often dreamed about drowning when I was a little girl.

I wasn't all that keen on exercise either; although I'd done a lot of rowing and running, it was mostly motivated by vanity rather than something I enjoyed for its own sake.

Where this idea came from, I will never know for sure. I like to think of it as an idea that wanted to happen, and pounced on me in a moment of vulnerability when my ego's guard was down. In my heart, I immediately knew it was exactly the project I'd been looking for. A split second later, my ego-mind weighed in, pointing out a hundred reasons why this was the worst idea ever. But over the course of a week, my mind gradually caught up with what my heart had known all along – that this was perfect for me.

From Peru to the Pacific

So it was that, four years after receiving that resounding call to adventure, I found myself in the middle of the Pacific Ocean. After rowing the Atlantic in 2005, I was continuing my mission to row the world's Big Three oceans, and the Pacific was next. After a failed attempt in 2007, this was now 2008 and I was trying again.

The Pacific Ocean is big. Really big.

This might be stating the obvious, but when I was planning to row it, I was quite shocked at just how mind-bogglingly big it is, so indulge me while I share some statistics. If I tell you it's about 64 million square miles, that's quite hard to imagine, so to put it a different way: if you took all the continents in the world and put them in the Pacific, you would still have enough room for another entire Africa.

And that's just the surface area. When you consider its volume, factoring in an average depth of 1 mile, bottoming out at 7 miles deep in the Mariana Trench, the Pacific is around 170 million cubic miles, just over half of all the oceanic water on the planet. This equates to over half the liveable volume of the Earth, because creatures can live throughout the water column, compared with the land, where most creatures live within a narrow band above and below ground level.

Enormous as the Pacific is, we have managed to pollute just about every last cubic inch of it with plastic.

You've probably heard of the Great Pacific Garbage Patch, which I've heard described as an island of floating trash the size of Texas. I've been asked many times if I saw it. No, I didn't. I'm not saying it's not there, but on my way from San Francisco to Hawai'i on the first leg of my Pacific crossing, I skirted around the edges of the Garbage Patch, and although I saw a few items of recognisable plastic, like a fishing net, mooring buoys and the occasional plastic bottle, I certainly didn't see anything you could describe as an island.

What I did see, however, is maybe more sinister. A few hundred miles east of Hawai'i, I met up with a couple of researchers on board an exceedingly homemade vessel – using the most generous interpretation of the word 'vessel' – made out of rubbish, and appropriately enough, called the Junk Raft.[2] Joel and Marcus showed me what they were finding in the manta trawl that they towed behind their boat periodically throughout the day to test how much plastic was in the water. The pieces they were finding were mostly tiny, just a few millimetres in size, invisible to the casual observer.

This may sound less horrific than the rumours I'd heard of floating TV sets, abandoned yachts and giant accumulations of ghost nets (discarded fishing nets), but the bad news is that these fragments are perfectly sized to make a handy snack for small fish, so they get into the oceanic food chain lower down, and accumulate to higher levels as they move up. As Joel and Marcus told me, the Great Pacific Garbage Patch is less of a patch, and more of a smog: diffuse clouds of plastic made brittle by exposure to sunlight and then pounded to pieces by waves. This makes it tremendously difficult to get the plastic out of the ocean without also taking out the good stuff, the plankton that form the bottom of the oceanic food pyramid.

In Hawai'i a few weeks later, I met up with the Hunks on the Junk (as the media had dubbed them) at a beach clean-up on Kahuku Beach on O'ahu's north-east coast. It is a rarely visited beach, so any litter that shows up there is unlikely to be left by day trippers and more likely to have been washed up from the ocean. I was shocked to see the amount of debris. Where I might expect to see lines of seaweed, instead I saw tidemarks of trash, most of it plastic. We took away about twenty bags of rubbish that day, but there was still plenty left lying on the sand, and an almost endless supply lurking offshore in the Garbage Patch, waiting to roll in on the next tide.

I was a frequent spokesperson on the plastic pollution issue around that time, not because I thought it was the most pressing issue, but because it seemed like a good gateway into environmental action. Climate change, for example, can be a tough sell. It's much easier for most people to picture how much disposable plastic they are using than it is for them to imagine a cloud of invisible greenhouse gases. And although I know they exist, I haven't yet knowingly met a plastic pollution denier. I was working on the assumption that a photograph of a tropical beach covered in plastic rubbish, or a turtle eating a plastic bag, or a seal ensnared in a packing strap, is more likely to evoke a visceral reaction – and hopefully action – than a hockey stick graph of carbon emissions.

I have to confess, then, that I fell into the common trap of addressing the problem that seems solvable, rather than the one that's the most important. I'm not saying that plastic pollution isn't important, nor am I saying that climate change should be the highest priority – I'm merely noting the human tendency to focus on cleaning up the elephant poop, rather than tackling the environmental elephant in the room.

The following year, after a winter of rest, recuperation and reprovisioning, I set out from Hawai'i and rowed the second leg of the Pacific, making landfall 104 days later in the Republic of Kiribati, a small island nation of about 100,000 souls, lying on both the equator and the International Date Line. It's the only country in the world that straddles all four hemispheres and consists of thirty-three low-lying islands, mostly less than 6 feet above sea level.

Understandably, its citizens are apprehensive about the future. Global mean sea level has risen between 8 and 9 inches (21–24 centimetres) since 1880, with about a third of that happening in the last twenty-five years. The rising water level is mostly due to thermal expansion of seawater as global temperatures rise, plus meltwater from ice caps and glaciers like the one I had seen in Peru.

Although 8 or 9 inches may not sound like much, when most of your country consists of coral atolls that barely protrude above sea level, it matters. When you consider that Kiribati has a land area of 313 square miles and 710 miles of coastline, it's clear that nowhere in the country is far from the rising tide.

Even before the atolls are inundated, increasing tropical storms are going to cause major problems, overreaching fringing reefs, contaminating fresh water supplies with seawater and washing away the fragile causeways that link the more inhabited islands. I met with President Anote Tong, who in our videoed conversation shared his fears for the future of his country.

I saw him again that December at the COP15 climate change conference, hosted by the United Nations in Copenhagen. It was the last Friday of the conference and he and his small delegation invited me to join them for dinner in an Indian restaurant. The conference had just ended in disappointment. The UN process had broken down, most of the national delegations had been excluded from the final rounds of talks (the Copenhagen Accord was drafted by only five countries: the United States, China, India, Brazil and South Africa), and there had been no fair and binding deal. While the conference had recognised that keeping average global temperature rise below 2°C would be a good thing, it made no commitments for reducing emissions in order to achieve the target, and had dropped proposals to limit temperature rises to 1.5°C and to cut CO_2 emissions by 80 per cent by 2050. The developed world pledged to help poor countries adapt to climate change, and offered to pay them to reduce emissions from deforestation and forest degradation (the REDD+ framework).

In effect, the failure of the conference had signed the death warrant for Kiribati and many of the other small island nations. In their speeches that night, the I-Kiribati contingent spoke brave words, but their despondency was palpable. 'We are trying to maintain our composure, but I am very sad,' President Anote Tong said to me. 'We were naïve and vulnerable. I wish I was so much more ruthless.'

I felt deep sadness for this dignified, elegant man, educated in New Zealand and with a master's degree from the London School of Economics. There was cruel irony in the fact that his studies of economics had not been able to prevent his country becoming collateral damage to neoliberal capitalism.

The following year, the president came down to the dock in Kiribati to wave me off on the third and final stage of my Pacific crossing. In his speech, he thanked me for my efforts in bringing attention to the plight of his country. As I rowed away, I reflected on the direness of their situation, and the pathetic insignificance of my endeavours on their behalf. I wondered how I would feel if it were my country destined to disappear under the waves, drowning the places where I had been born, grown up, gone to school, and where my ancestors were buried. I couldn't help thinking that if their economy was worth $15 trillion (US GDP for 2010) rather than $150 million (Kiribati GDP for 2010), their future would be taken much more seriously by the global community. As it was, this proud island nation was apparently seen as dispensable.

A Crisis of Conferences

Immediately before I set out on that last leg of my voyage, from Kiribati to Papua New Guinea, I spoke at a TED conference on board the *National Geographic Endeavour* in the Galapagos Islands, in honour of the renowned American marine biologist Sylvia Earle and her TED Prize-winning wish to protect our oceans.

On the one hand, the conference was good for my ego, rubbing shoulders with a star-studded audience including A-list celebrities like Leonardo DiCaprio, Edward Norton, Glenn Close, Chevy Chase, Damien Rice and Jackson Browne. On the other hand, it was bloody depressing. Species extinction, ocean acidification, collapsing fish stocks, coral reef destruction, fossil fuel extraction, sewage, agricultural run-off, sea dumping, mining, plastic pollution, toxic pollutants like PCBs and DDT – amidst all these attacks and acronyms there seemed to be precious little hope for the survival of the oceans, and hence the survival of anything else on this planet. Jeremy Jackson, a gaunt marine ecologist rejoicing in the moniker of Doctor Doom, put the final nail in the coffin of my morale with his talk, entitled 'How We Wrecked The Oceans'. It made me want to curl up in despair rather than keep rowing.

But keep rowing I did, and after a total of 250 days at sea since leaving San Francisco, I completed my Pacific crossing in Madang, Papua New Guinea. The official estimate was that 5,000 people turned out to greet me. I can only vouch for the fact that there were a lot, including many in decorated canoes and traditional costumes, others waving homemade banners bearing my name, and I was given more gifts than I could carry. I was overwhelmed by the warmth of the welcome.

The day after I arrived, there was a major demonstration in the capital, Port Moresby, to protest against government corruption. The big news in Madang, meanwhile, was the ongoing controversy over a Chinese-owned nickel mine that was being constructed in the jungle nearby. There were three big counts against the mine: concerns that the tailings would damage the reefs of the Coral Triangle; the mine being staffed entirely by Chinese, with no jobs for the local people; and the mining company being given a twenty-five-year tax break by the Papua New Guinea government. These complaints, particularly the last, seemed not entirely unrelated to the anti-corruption protests.

During my time in Madang I also learned of families selling their ancestral land to logging companies, sacrificing their family's future in favour of present profit, and of the widespread exploitation of Papua New Guinea's rich natural resources by foreign companies taking advantage of this relatively new and chaotic country, with its 851 languages and a largely rural, tribal population. Papua New Guinea is one of the world's least well-documented countries, but is believed to have numerous uncontacted tribes, and many species of plants and animals as yet undiscovered. I could almost sense greedy first-world corporations drooling in anticipation as they contemplated its natural riches, corruptible politicians, financially unsophisticated tribespeople and inadequate environmental protections.

Over the course of the years, as many ocean miles passed under the hull of my rowboat and I continued to show up for conferences and marches, it was dawning on me that there was much more to solving this ecological crisis than simply pointing out to decent folks that we have a problem. Even the conferences themselves, with the notable exception of TED Mission Blue in the Galapagos Islands, often seemed abstracted, transactional and businesslike, with few reminders in the conference halls of the precious natural world we were supposedly trying to protect. Presentations often centred around scientific data, cost-benefit analyses and business cases. There was a lot of head and precious little heart – the dynamic that created the problems in the first place.

Ecological damage has been generated primarily by the more developed world, and suffered primarily by the less developed world. There is a geographical disconnect, compounded by wilful blindness. As the environmental journalist George Monbiot said at a talk I attended in Copenhagen, the attitude of the Global North seems to be: 'If I swing my fist and my neighbour's nose happens to get in the way ... tough.' The economic disparity between the Global North and the Global South allows the North to continue its exploitation, and affords it the luxury of insulating itself from the worst impacts.

This colonialist attitude is written into every climate change agreement there's ever been. Look at the Copenhagen pledges as an example: the developed world pledged to help poor countries adapt to climate change. Translates to: 'We've screwed your country – oops, sorry, here's some money.' They also offered to pay developing countries to reduce emissions from deforestation and degradation. Translates to: 'We've deforested and degraded your country for the last couple of centuries, but we don't want you to do the same. Here – have some more money.'

While in Copenhagen I spent time with a new friend, the Scottish barrister and author of *Eradicating Ecocide*, Polly Higgins. She vehemently opposed the monetary valuation of nature, seeing it as sacred in its own right, regardless of what it delivers to humans in ecosystem services. Polly was an inspiring visionary and big-hearted human being. She died too soon, aged 50, in 2019, less than a month after being diagnosed with an aggressive cancer. She lives on, though, in the ideas she sowed and the campaigns she started. In particular, she could see that corporations would continue to exploit our planet while the only penalty they faced was a financial slap on the wrist. Her dream was to have ecocide recognised as an international crime, alongside genocide, crimes against humanity, war crimes and crimes of aggression, for which CEOs could be prosecuted and sent to prison. Her dream hasn't yet come true, but the work continues.

It seemed to me increasingly clear that we were trying to solve the environmental problems from the level of economic consciousness that created them. The prevailing attitude seemed to be that everything had its price, from trees to oceans to people. If you had enough money to throw at the problem, you could pay off the poor, bribe politicians and spend your way out of environmental legislation as an acceptable cost of doing business.

Some things are precious beyond price, and yet I went to ocean conference after environmental conference where the talk was all about money – money for research, money for conservation, money for mitigation and adaptation and compensation. When I went to a conference hosted by *The Economist* (which should have been a clue) and the discussion turned to coral reef insurance, I felt sick. It was becoming increasingly clear to me that we couldn't fix our ecology without first fixing our economy.

Chapter 2

Economics

'A stark choice faces humanity: save the planet and ditch capitalism, or save capitalism and ditch the planet.'[1]

Fawzi Ibrahim

In 2018 I was invited to the Mount Washington Hotel in Bretton Woods, New Hampshire, for a conference called the Global Economic Visioning. This was the same hotel – a grand, turreted edifice, white-walled with a bright red roof, set against the dramatic backdrop of Mount Washington, the highest mountain in the north-eastern US – where the original Bretton Woods conference had taken place in 1944.

For a while, I'd been interested in economics as a leverage point for environmental change – or, conversely, as the rock on which such change too often foundered – but realised I didn't know much about the first Bretton Woods, or, as it was officially called, the United Nations Monetary and Financial Conference. It was time for some homework.

Towards the end of the Second World War, for three weeks in July 1944, 730 delegates (all men) from the forty-four Allied Nations convened at Mount Washington. The most pressing issue of the day was how to prevent another world war, this one having arisen, at least in part, out of the economic misery in Germany resulting from post-war reparations and compounded by the Great Depression, providing fertile ground for fascist ideas to take root.

The British delegate, John Maynard Keynes, was the key figure at the conference. He argued that the rise of protectionism (the imposition of tariffs on imported goods) across the industrialised world had contributed to low aggregate demand: people weren't buying enough stuff because prices were too high, and the economy had ground to a halt. For Keynes, excess protectionism was a primary driver of the Great Depression, so the goal of the new General Agreement on Tariffs and Trade (GATT) was to reduce tariffs across the board in order to boost demand.

In other words, the GATT began its life as an ostensibly benevolent institution, intended to promote economic stability and peace. Just like the original International Monetary Fund (IMF) and the World Bank, founded during the same conference, it was rooted in Keynesian principles of full employment and economic stability, and intended to contribute to the collective good.

For many years the Bretton Woods Agreement worked well, but things started to go awry in 1971 when Richard Nixon unilaterally withdrew the US from the gold standard, meaning that the dollar no longer represented something with real value – actual gold – and instead became bits of paper and cheap metal that had value only because people *believed* they had value. Currency became a fiction that worked only because everybody agreed to believe the same story.

Before 1971, the US could only print as much money as was backed by its gold reserves, but now it could print as much money as it liked. And because the US dollar was the global reserve currency, meaning it was the default currency for international transactions, this put the US at a huge advantage. For any other country to obtain US$100, it had to produce goods or services worth $100 and sell them. For the US to obtain $100, it simply printed them.

In the 1980s, further blows were dealt to the spirit of the Bretton Woods Agreement. The economic stability that had been a key objective depended on tight financial regulation, but under Margaret Thatcher and Ronald Reagan regulations were relaxed or removed altogether. Free trade reforms dismantled many of the capital controls that were intended to protect the stability of national economies. Investors could now withdraw their capital at short notice from a particular country if they thought they could make more money elsewhere, and if several investors did this at once, it could devastate a country's economy.

For several decades, there had been increasingly loud calls for a new Bretton Woods, particularly to create a new international framework to tackle several pressing issues: the unfettered capital flows that were creating global financial crises; a far from level playing field in terms of each nation's access to capital; and the exclusion of many developing countries from economic progress and political participation.

The conference I attended was interesting, with an outstanding opening session featuring two of my favourite thinkers, Charles Eisenstein and Daniel Schmachtenberger, but the new Bretton Woods it wasn't. There was a lot of excitement about cryptocurrencies, some of which have potential, but most of which are being used for old-school, get-rich-quick speculation, just like any other commodity. Cryptocurrencies used this way may as well be gold, oil or pork bellies. To be fair, the conference didn't set out to overhaul the entire global economic model – its stated goal was to update the conversation, incorporating newfangled, post-1944 developments like the internet, crypto and the radical idea that women might have something valuable to contribute to the debate.

My preparation for the conference, and my frustration at its limitations, helped me clarify my personal view on what we need in the way of economic reform. My grade A in A-Level economics hardly qualifies me to pontificate on the state of global capitalism, but from my simplistic perspective, some things seem quite obvious. I believe we need to step outside our current story of money, and take a long hard look at our economic system from the outside. We created capitalism, and while it has brought many benefits, it is also encouraging the destruction of our ecosystems, and widening the gulfs of inequality both within and between nations.

There are many signs that the old story of money isn't working – the dubious correlation between money and happiness, how we've been brainwashed into buying stuff we don't need, the non-stop accumulation of wealth and debt, externalisation of (environmental) costs, corporate lobbying, misinformation campaigns, business-sponsored science, short-termism and much more besides. These are huge topics, and by necessity I'm only skimming the surface of a few selected themes.

To be clear, I have nothing against money, capitalism, currencies, cryptocurrencies, economics, economists or anything else in this arena. I'm in favour of anything, so long as it delivers the greatest good to the greatest number over the longest possible period of time. If a story doesn't contribute to that objective – recalling that money is just a story – then it's worth seeing if we can come up with a more effective one.

My sense is that we need to come up with a vision of the world we want to create – for example, a world of peace, sustainability and fairness – and reverse engineer from there to figure out what principles and structures will support that vision. The system we have now isn't working – not for the planet, not for the developing world, and not for 99 per cent of the people in the developed world. We can do better than this.

Metrics of Success

In 2012, I was fortunate enough to be one of sixteen international Fellows selected for the Yale World Fellowship Programme (of which more later). One of the other Fellows tipped me off that Open Society International was offering grants to fund transformative projects. Fired up with world-changing zeal by my time on the programme, and otherwise in a between-roles phase of my life, I decided to give it a shot.

I prepared an application for a project I called the New Prosperity Paradigm. I had come to believe we need a new definition of success, accompanied by a new metric to replace gross domestic product (GDP). The evidence, as I saw it, suggests that our current definition of success, based on fame, fortune or position, is not enhancing happiness and wellbeing, but instead is delivering material aspirations that run counter to the wellbeing of the planet, as well as to the wellbeing of humans.

This was my central premise: up to a certain point, yet to be achieved by many developing countries, rising income undoubtedly correlates to rising wellbeing. But once basic needs are met, the marginal benefits of additional income rapidly diminish – but consumption does not. Additional purchasing power doesn't yield a proportionate increase in current wellbeing, and in terms of resource extraction and environmental pollution, it seriously impacts on the potential wellbeing of future generations.

Ultimately, my grant application failed in the sense that I didn't get my project funded. But it succeeded in that I learned a lot about the global economy and the way it's exploiting the vast majority of humanity, as well as causing enormous damage to the natural world. I wouldn't mind so much that we are trashing our Earth if it was making us happier, but it's not – not even the rich folks.

The conventional approach is based on a faulty assumption: that the best way to measure improvement in living standards of a country is by looking at the growth rate of its gross domestic product per capita, which refers to the value of goods and services produced within a country's borders. But does rising GDP cause a widespread rise in the standard of living? And does *standard of living* equal *quality of life*? It would seem not.

As Robert F. Kennedy famously said of GDP in his campaign speech in 1968, 'It measures everything, in short, except that which makes life worthwhile.'[2] Even the creator of the GDP metric, Simon Kuznets, said in 1934, 'The welfare of a nation can scarcely be inferred from a measurement of national income.'[3] A quick online search yields dozens of quotes about the inadequacy of GDP (and also of its close cousin, GNP, gross national product, which is the value of goods and services produced by a country's citizens both domestically and abroad). Those critics range from Ban Ki-Moon, former UN secretary-general, to the late Robert McNamara, former president of the World Bank.[4] In *Doughnut Economics*, Kate Raworth points out that we have been fixated on GDP for over seventy years, and that this fixation has been used to justify extreme inequality and the unbridled destruction of the natural world. She advocates for a new goal – ensuring that human rights are met for all, within the limits of our planetary boundaries.[5]

GDP has been weaponised by multinational corporations in the oppression and exploitation of less developed countries, as John Perkins describes in *Confessions of an Economic Hit Man*. His job was to convince foreign governments to embark on massive infrastructure projects 'intended to create large profits for the contractors, and to make a handful of wealthy and influential families in the receiving countries very happy, while assuring the long-term financial dependence and therefore the political loyalty of governments around the world. The larger the loan, the better.'[6] Why would foreign governments agree to such exploitative contracts? They were persuaded by outrageously optimistic promises of GDP growth, which would make them look good to their citizens and to their peers on the global stage.

It's not only leaders of less developed countries that are susceptible to GDP vanity. Professor Tim Jackson writes, speaks and advises extensively on prosperity without growth. He tells the story of being summoned to meet with a senior adviser at the UK Treasury, where he was invited to set out his stall: he spoke of how our addiction to ever-increasing GDP is destroying the planet; the challenges of decoupling economic growth from unsustainable practices; and the necessity of supplanting consumerism with a different conception of prosperity based on genuine wellbeing. The adviser listened carefully, then asked, 'What would it be like for Treasury officials to turn up at G7 meetings knowing that the UK's GDP had slipped down the rankings?'

Jackson says he was 'dumbstruck' by this response. 'How could I have missed that the politics of the playground is evidently still in action, even in the highest echelons of power?'[7] Apparently, the preservation of the planet is less important than a Treasury official's ego.

The trouble is that GDP deceives us with its reductionist simplicity. In a world that finds it preferable to measure quantities rather than qualities, it pretends to be a proxy for human thriving. We know it's a crude metric – it includes as positives many things that are clearly antithetical to wellbeing, such as the cost of wars, cancer care, rebuilding houses destroyed by wildfires, and depression medications, while ignoring the beautiful immeasurables of love, joy, fulfilment, relationship and laughter.

And yet we still use it. It takes the glorious messiness of human society and boils it down to a single number, enabling a government to show that the number is getting bigger under its leadership, or that its number is bigger than another country's number. It's a blunt instrument used to bludgeon us into believing that things are getting better, when mostly they aren't.

Incentives

If GDP is the measure of a nation's success, then profits are the measure of a corporation's success, and both metrics suffer from many of the same problems. My fundamental issue with laissez-faire (aka free-for-all) capitalism is its incentives. As billionaire investor Charlie Munger says, 'Never, ever, think about something else when you should be thinking about the power of incentives.'[8] Every animal, including humans, is designed to seek rewards, and will adjust its behaviour accordingly: a dog for a biscuit, a dolphin for a fish, a banker for a bonus.

While profit maximisation is the preeminent goal, so far ahead of any other goal that the others have just about disappeared over the horizon, executives are incentivised to overlook minor little things like destroying the Earth and exploiting humans as if they were just another resource to be used up and worn out. They are rewarded for perpetually feeding and growing the bottom line, even if that is best served by increasing misery, waging war or exploiting natural disasters. John Maynard Keynes summed it up, 'Capitalism is the astounding belief that the most wickedest of men will do the most wickedest of things for the greatest good of everyone.'[9]

This isn't personal. Many executives are good people, and many companies do good things. I'm just saying that a *system* that prioritises profit over all else encourages the people who work within that system to set aside morals or scruples if there is a fast buck to be had. If what you measure is what you get, then measuring profit but neglecting principles will inevitably lead to unprincipled behaviour.

Money Doesn't Buy Happiness

When I say that money doesn't buy happiness, that's not to say that money doesn't matter. Clearly, a person earning $70,000 a year is going to have a dramatically better quality of life than a person living below the official International Poverty Line of about $700 a year. At this level, the evidence unsurprisingly tells us that yes, richer people are happier.[10] At an individual level, there are of course happy poor people and miserable rich people, but in general it's a lot easier to be happy if you have access to decent food, clean water, shelter, education and healthcare.

According to some research, the correlation between income and well-being breaks down above $75,000 due to hedonic adaptation – essentially meaning that too much is never enough.[11] Then again, another study says that to achieve 'optimal contentment' each of us would need to possess $100 million.[12] If that's really what it takes, most of us will have to get used to *sub*optimal contentment.

So, opinions vary. But a useful question here might be: to what extent is wealth a determinant of happiness, or is happiness a state of mind? J. Paul Getty was the richest man in the world, but was also a miserable sod, according to David Scarpa, the screenwriter of *All the Money in the World*, the 2017 film about Getty's refusal to pay the ransom for the return of his grandson. The kidnapping took place during the oil crisis of 1973, when Getty was making enough profit *every day* to more than cover the ransom, but he was so hooked on increasing his wealth that he refused to pay. 'The wealthier he became,' says Scarpa, 'the more dependent he became on money, like an addict.'[13]

Or, as Charles Eisenstein put it when he appeared on my podcast in 2021:

> 'The remedy is to understand that the deep, authentic human needs that drive endless acquisition and greed are tragically unmet in our society. The need to belong, the need for connection, the need for intimacy, the need to know the faces and places around you, to be embedded in a web of stories, in relationships, the need to be at home in this world, cannot be met by money. And if you try to meet it with more and more money, how much more money will be enough? No amount will be enough. So it's like we have a society of winners and losers. And the tragic irony is that even the winners are losers.'

So we have to question whether more is always more, or if there might be a kind of *terminal happiness*, like terminal velocity, beyond which more money can't buy you any more. Sonja Lyubomirsky, professor of psychology at the University of California, Riverside, proposes that people have a happiness set point that is 50 per cent inherited and 40 per cent due to intentional activity, meaning our life choices about how we spend our time and energy. Only 10 per cent is due to life circumstances, including how much money we do or don't have. A study on adaptation level theory tested twenty-two major lottery winners, and noted that their win didn't make them as happy as we (or they) might expect them to be. Winning the lottery had been such an exciting experience that it diminished the joy of simple pleasures, and once the novelty had worn off, everything felt a bit flat.[14]

It seems likely that, happiness-wise, we fall on a standard distribution curve, with a small percentage of people at the depressive end of the scale, a small percentage at the deliriously happy end, and the vast majority of us getting along pretty well in the middle *irrespective of how much money we have*, provided our basic needs are met.

Could there even be ways in which a simple life is more conducive to wellbeing? I don't for a moment want to idealise poverty, but I'm remembering a time in 2010 when I was sailing with a group working on coral reef observation and preservation in Papua New Guinea, and on the evening in question our boat was in the harbour of a small island. The people there had no electricity, extremely limited infrastructure and a diet consisting of the narrow range of foods that could be foraged from the forest or the sea. Yet, as night fell, the forest echoed to the sound of laughter, a sound you would rarely hear in even the most affluent of London streets. Was this just an anomaly, or could we be wrong in assuming a direct relationship between money and happiness?

For her book, *Happier People, Healthier Planet*, Dr Teresa Belton interviewed dozens of people in the UK who had chosen to live simpler lives. Echoing Charles Eisenstein, her conclusion was:

> 'Consumption patterns in the developed world, and to which many in the developing world aspire, are doing irreversible damage to the global climate and other aspects of the environment, yet they are entirely unnecessary for leading a happy and satisfying life ... It is not material wealth but non-material assets, such as strong relationships, active engagement, and thriving communities, that enrich our personal and social wellbeing.'[15]

So whether these people are less materialistic because they are already happy, or whether they are happy because they are less materialistic, it's worth questioning the causal link between affluence and wellbeing.

Even if you believe that money *can* buy happiness, wouldn't we all agree that the economy should be a lower priority than preserving a liveable planet? After all, if we alter our ecosystems to the extent that vast numbers of humans are being wiped out by heatwaves, famine and water shortages, doesn't the state of the economy become somewhat irrelevant? If you have one iota of humanity in your heart, no amount of money can make you happy in a world like that.

Brainwashed into Buying

So why do we keep buying so much stuff? Tim Jackson pithily sums up the insanity of debt-fuelled consumerism in his TED Talk:

> 'This is a strange, rather perverse, story about people being persuaded to spend money we don't have on things we don't need to create impressions that won't last on people we don't care about.'[16]

The Century of the Self is a compelling documentary series by Adam Curtis (available on YouTube), which focuses on the so-called grandfather of public relations, Edward Bernays, and the extent to which people's desires have been manipulated by marketing strategies. It quotes Paul Mazur, who was a banker with Lehman Brothers in 1927:

> 'We must shift America from a needs to a desires culture. People must be trained to desire, to want new things, even before the old have been entirely consumed ... Man's desires must overshadow his needs.'

Bernays' book, tellingly entitled *Propaganda*, drew on the work of his uncle, Sigmund Freud, and described how the intelligent and powerful could – and, in his view, should – manipulate the masses. This quote is particularly chilling:

> 'The conscious and intelligent manipulation of the organized habits and opinions of the masses is an important element in democratic society. Those who manipulate this unseen mechanism of society constitute an invisible government which is the true ruling power of our country. We are governed, our minds are moulded, our tastes formed, our ideas suggested, largely by men we have never heard of.'[17]

A case study: Bernays was hired just after the First World War by the CEO of the American Tobacco Company to help figure out how ATB could tap into the female half of the market. 'If I can crack that market, it will be like opening a new gold mine right in our front yard,' the CEO apparently said.

At the time, it was deemed unseemly for women to smoke in public. They were liable to be branded whores or harlots. A major cultural shift in public perception was needed. No doubt highly motivated by his $25,000 incentive, a vast sum at the time, Bernays came up with possibly the most audacious rebranding campaign in history.

On 31st March 1929, at the height of New York's Easter Parade, a young woman named Bertha Hunt stepped out into the crowded Fifth Avenue and lit a Lucky Strike. The press had been tipped off in advance and given leaflets containing relevant information (propaganda). What they didn't know was that Hunt was Bernays' secretary and that this was the first in a series of events aimed at getting women to smoke.

Hot on the heels of American women getting the vote in 1920, Bernays proclaimed to the press that smoking was a form of liberation for women, a chance to express their strength and freedom. The great irony was that he was using promises of liberation as a form of control; what he branded *torches of freedom* actually enslaved countless women into nicotine addiction. But the capitalist objective was served.

I wish I could believe that this toxic and manipulative attitude died out with Edward Bernays, but a contemporary advertising executive is also quoted in the documentary: 'It's our job to make people believe they suck if they don't have our product.'

As John Perkins documents in his account of economic hitmen, there are still those who see themselves as powerful puppet masters. He fondly imagines that awareness will lead to transformation. I wish I shared his optimism, but these are powerful forces.

> 'I am certain that when enough of us become aware of how we are being exploited by the economic engine that creates an insatiable appetite for the world's resources, and results in systems that foster slavery, we will no longer tolerate it. We will reassess our role in a world where a few swim in riches and the majority drown in poverty, pollution, and violence. We will commit ourselves to navigating a course toward compassion, democracy, and social justice for all.'[18]

The rise of the advertising industry coincided (possibly not a coincidental coincidence) with the decline in religion. As the divine retreated from Western life, a void arose, which advertising rushed to fill. Advertisers dupe us into believing that the thing we lack just happens to be the thing they're selling. When we try to plug that existential black hole with a new car, or face cream, or electronic gizmo, the new object of our delight might momentarily distract us from the black hole, but it doesn't fill it – it simply disappears into the hole's depths. We can keep buying and buying and buying – if we can afford it, we could spend an entire lifetime distracting ourselves with material purchases – but the hole will keep on gobbling up our goodies, because what we're yearning for isn't material at all. While we're desperately trying to plug the *hole*, we'd be better off looking at the *whole*; in other words, focusing on becoming the whole, actualised person we were born to be.

Money and the Great Mother

So where did it all go wrong? Surely we invented the economy to serve humanity. So when did humanity start being in service to the economy?

Bernard Lietaer, the wonderful and wise Belgian economist who sadly was lost to this world in 2019, took a perspective based on a reinterpretation of Jungian psychology. The Swiss psychiatrist, Carl Gustav Jung, believed that the collective human unconscious contains a pantheon of characters, or archetypes, which influence how we experience our lives. Any compelling story incorporates archetypal figures such as the Warrior (Jon Snow, Luke Skywalker), the Magician (Bran Stark, Yoda), and the Queen/King (Daenerys Targaryen, Darth Vader – although he's not the kind of king you'd turn out and wave a flag for).

Lietaer maintained that much of the trouble that we have with money – gross inequality of distribution, boom and bust, metrics that encourage exploitation rather than conservation – is caused by the absence of the benevolent, nurturing, fiercely protective Great Mother archetype from our money systems, and, in fact, from our world at large. (Cultural examples of the Great Mother are hard to find, which rather proves Lietaer's point. Think of Mother Nature, or Forrest Gump's mother, but a thousand times more so, and not having sex with the school principal.)

In Lietaer's view, the crises of ecology, community breakdown and violence can only be resolved through the reintegration of the feminine perspective into our worldview. While the unhealthy variant of masculine power continues to predominate, so will our problems.

He tracks the history of the repression of the Great Mother over the centuries since the third millennium BCE, by the Mesopotamians, the Greeks (Aristotle wrote, 'For the female is, as it were, a mutilated male'; I'd like to have words with Aristotle), the religious scriptures of Judaism and Christianity, and the witch hunts of the Middle Ages.

The problem arises when an archetype is suppressed, because it doesn't just go away. Instead, the society expresses the archetype's shadow, or negative, aspects instead. Lietaer presented a model in which each archetype has a yang/masculine shadow, and a yin/feminine shadow, in which the shadows have too much or too little of the archetypal quality.

According to ancient Chinese philosophy, as discussed in the three-millennia-old text, the *I Ching*, the universe is governed, even created, by yin and yang, two opposing yet complementary principles. Generally, yin is characterised as the feminine principle (inward-looking, dark and still, representing the negative polarity) while yang is the masculine principle (outward-looking, hot and bright, representing the positive polarity).

We can see how these yin- and yang-flavoured shadows show up by giving some examples: the shadow sides of the Warrior archetype are sadism (yang) and masochism (yin). The shadows of the Lover are addiction (yang) and impotence (yin). The shadows of the King/Queen are tyrant (yang) and abdicator (yin). And the yang and yin shadows of the Great Mother are greed and scarcity.

When you look at our financial systems, do you see the nurturing, abundant, loving influence of the Great Mother? Or do you see greed and scarcity? I know what I see, and it's not at all motherly. Let's look at scarcity first.

In *Rethinking Money*, which Bernard Lietaer wrote with Jacqui Dunne, the authors are not condemning money as such. They point out that the emergence of the financial system 'produced remarkable advances, thrusting society out of the shackles of superstition and stagnant social order that had preceded it. It brought about the rigour of science founded in that which could be proven, rather than divine dogma. It enabled the individual, no matter how lowly his birth, to scale the heights of his unbridled imagination and keen ambition through learning and labour.'[19]

Rather, they are saying money needs a rethink, as per the title. In relation to scarcity, they reference economics professor L. Randall Wray's assertion that governments choose what they will accept as payment of taxes from the citizens, and this determines what counts as 'money'. In modern economies, governments invariably choose the national currency – dollars, pounds, yen and so on – but they could have chosen seashells, pine cones or toenail clippings. The citizens then have to find ways to obtain the wherewithal to pay their taxes, and if they're not able to collect it from beaches, woods or the bathroom floor, they have to sell their time or their goods in order to raise the money to pay taxes in a form acceptable to government. Money is not a natural phenomenon – it is a means chosen by government, and assented to by the population.[20]

In other words, a government doesn't *need* to levy taxes to pay for its expenses. It can simply print more money. As Ezra Pound said, 'To say that a state cannot pursue its aims because there is no money, is like saying that an engineer cannot build roads, because there are no kilometres.'[21]

A government has to limit the money supply because money needs to have a degree of scarcity in order to maintain its value. According to the laws of supply and demand, if there is too much money sloshing around the system, each dollar becomes worth less, and hence the taxes paid by the citizenry to the government lose value.

Scarcity is also generated by the interest charged on loans. Lietaer and Dunne compare this to a game of musical chairs in which the bank-debt monetary system chooses not to create enough money to pay the interest on all the loans, so – as with the missing chair – someone ends up without. This manufactured scarcity in turn creates a never-ending push for growth, because borrowers must find additional money to pay back the interest on their debt.

So, essentially, it is taxes and interest-bearing debt that create scarcity. It is important to note that scarcity is not a natural law of the universe, or even of money. Daniel Defoe wrote in 1726, 'Things as certain as death and taxes, can be more firmly believ'd'. But of those two, only death is non-negotiable. As for interest, until the thirteenth century it was illegal. According to the Qur'an, it still is.

When it comes to greed, recall Gordon Gekko's classic speech in *Wall Street*, in which he lauds greed as being the ultimate driver of progress and the saviour of the United States. The Gekko character was intended to be a villain, a cut-throat, self-centred, ruthless scumbag. However, for some he became a role model and the instigator of many a trader's career.

John Perkins, in *Confessions of an Economic Hit Man*, argues that greed is not at all good:

> 'When men and women are rewarded for greed, greed becomes a corrupting motivator. When we equate the gluttonous consumption of the earth's resources with a status approaching sainthood, when we teach our children to emulate people who live unbalanced lives, and when we define huge sections of the population as subservient to an elite minority, we ask for trouble. And we get it.'[22]

Wealth on One Side, Debt on the Other

The current form of capitalism has benefited the few at the expense of the many. In *The Divide*, Eswatini-born Jason Hickel contends that the widespread poverty that has plagued Africa, South America and large parts of Asia is not due to any lack of the ability of its people to farm and govern effectively, but due to the rapacious behaviour of Europeans and North Americans over the last 500 years. After the conquistadors and slave traders came the governments and corporations that have conspired (in the most criminal sense of the word) to keep the South disempowered and downtrodden.

Jason Hickel describes how his suspicions were first aroused when he noticed that AIDS patients were dying, not because of lack of availability of medication, but because pharmaceutical companies jealously guarded their patents and wouldn't allow Eswatini to import generic versions of life-saving drugs. He saw that farmers couldn't make a living off the land because subsidised foods were pouring in from the US and the EU, undercutting their prices. He observed that the government was being forced by international banks to prioritise interest payments on foreign debt over providing basic social services to its citizens. He grew incensed by the human suffering caused by heartless capitalism.

Jason Hickel has attracted more than his fair share of critics. This may be less to do with his research and assertions, more to do with the discomfort that his views generate in those who have benefited from the paradigm he criticises. As the political activist Upton Sinclair used to say, it's difficult to get a man to understand something when his salary (or his standard of living) depends upon his not understanding it.

In the 1990s, the developing world was spending $13 on debt repayment for every dollar it received in foreign aid and grants. By 2004, that number had grown to $20 on debt repayment for each dollar of foreign aid. Former president of Nigeria Olusegun Obasanjo reflected ruefully on his nation's debts:

> 'All that we had borrowed up to 1985 or 1986 was around $5 billion and we have paid back so far about $16 billion. Yet, we are being told that we still owe about $28 billion. That $28 billion came about because of the foreign creditors' interest rates. If you ask me, "What is the worst thing in the world?" I will say, "It is compound interest."'[24]

To add insult to injury, as mentioned in the previous chapter the developed world is now pledging money to help the less developed nations adapt to the impacts of climate change – with the money coming *in the form of loans.* So, stated bluntly, the developed countries produced most of the historical carbon emissions, before air pollution and the quest for cheap labour drove them to outsource their dirty industries to developing countries. Most developed countries have temperate climates, so are less impacted by climate change, while equatorial developing countries face the most severe consequences – consequences they played little part in creating.

To add yet further insult, according to George Monbiot, writing in *The Guardian* during the COP26 climate change conference, since 2009 the rich countries have promised $100 billion to poorer countries in the form of climate finance, a paltry sum compared with the $3.3 trillion (yes, trillion) that they have spent on subsidising fossil fuel industries, and a fraction of the sums they have spent pulling up the drawbridges to immigration. Between 2013 and 2018, the UK spent almost twice as much on keeping migrants out as it did on climate finance. For the US the multiple was eleven, Australia thirteen, and Canada spent fifteen times more on border protection than on supporting the victims of our addiction to materialism.

Looking at this from the perspective of a citizen of the developing world, I would be feeling extremely indignant by now. You mess up the climate with carbon dioxide. You mess up my country with factories. You make me work for slave wages. Now my country is beset by storms, floods, failing crops and famine. You won't let me into your country. And now you want to make me pay for the privilege of borrowing money so I can try and keep my family safe and fed – really?

The more a country struggles under the impacts of climate change, the harder it will be to pay the interest on the loans. And so the economic colonisation by the rich over the poor (and predominantly white over brown and black) continues.

When I read Jason Hickel's tales about democratically elected leaders of developing countries being removed by coup or by assassination, orchestrated from the Global North, and replaced with dictators friendly to northern economic agendas, I was gobsmacked. Bumping off rulers you don't like sounded like a plotline from *Game of Thrones*. Surely such skulduggery could not be possible in the modern world?

But economic hitman John Perkins corroborates these stories. Omar Torrijos, at the time the de facto leader of Panama, and Jaime Roldós Aguilera, president of Ecuador, both expressed a fierce desire to protect the rights of their people against foreign exploitation. By *extraordinary* coincidence, they both met violent ends while still in office, in highly suspicious plane crashes.

But hang on, you might be thinking. Doesn't the Global North send lots of aid overseas? Surely we're the good guys?

Hickel quotes a study by the US-based Global Financial Integrity (GFI) and the Centre for Applied Research at the Norwegian School of Economics. They tallied up all the financial resources that get transferred between rich and poor countries each year: not just aid, foreign investment and trade flows, but also other transfers like capital flight, debt cancellation and remittances. They found that in 2012, the last year of recorded data at the time of the study, developing countries received a little over $2 trillion, including all aid, investment and income from abroad. But more than twice that amount, around $5 trillion, flowed out of them in the same year. Since 1980, developing countries have paid $4.2 trillion in interest payments, far more than they received in aid during the same period.

But, you may say, capitalism itself is not bad. It has just been abused by bad people. The proof is in the pudding of the great neoliberal experiment. Hasn't it alleviated poverty and hunger? Hasn't it made the world a better place?

According to Hickel, no.

In 1974, at the first UN Food Conference in Rome, US Secretary of State Henry Kissinger promised that hunger would be eradicated within a decade. At the time, there were an estimated 460 million hungry people. Today there are about 800 million hungry people, even according to the most conservative measures. More realistic estimates put the figure at around 2 billion – nearly a quarter of all humanity.

The poverty headcount is exactly the same now as it was when measurements began back in 1981, at about 1 billion. There has been no improvement during the last thirty-five years. In reality, the picture is even worse. The standard poverty measure counts the number of people who live on less than a dollar a day. But in many Global South countries a dollar a day is not adequate for human existence, let alone dignity. Many scholars are now saying that people need closer to $5 a day in order to have a decent shot at surviving until their fifth birthday, having enough food to eat and reaching normal life expectancy. If we measured global poverty at this more realistic level, we would see a total poverty headcount of about 4.3 billion people. Any reported alleviation in poverty has been mostly attributable to the World Bank fudging the figures, according to Hickel. If you take China and East Asia out of the statistics, it is even more starkly clear that poverty has worsened.

So why the deception? Why bother to pretend that hunger and poverty are decreasing, when they're not?

It's because those who benefit from neoliberal capitalism *have* to pretend that it is working, and working not just for them, but for everybody. They don't want to admit to the world – maybe not even to themselves – that their good fortune depends upon, and always will depend upon, the suffering and oppression of countless others.

But doesn't a system that rewards initiative and entrepreneurship mean more wealth for everybody? This was an argument favoured for a while by those at the top of the pyramid. But even Warren Buffett knows that trickle-down economics doesn't work. In 2018, when he was the third richest person in the world with a net worth of more than $86 billion, he admitted that many individuals suffer even as those at the top prosper wildly.[25] He points to the Forbes 400, which lists the wealthiest Americans. In his words: 'Between the first computation in 1982 and today, the wealth of the 400 increased 29-fold – from $93 billion to $2.7 trillion – while many millions of hardworking citizens remained stuck on an economic treadmill. During this period, the tsunami of wealth didn't trickle down. It surged upward.'

Thomas Piketty, author of *Capital in the Twenty-First Century*, describes how the current dynamics make the rich richer. When you acquire or inherit a sum of money that far exceeds what you need to live on, you can afford to make risky investments, which offer a significantly better rate of return, on average, than more conservative investments. The richer you are, the more extravagantly you're able to gamble, and the richer you become. In the US, the government has further helped those who least need their help, after the uber rich funded a successful lobbying campaign to dramatically raise the threshold for inheritance tax, further contributing to the accumulation of inherited wealth. Plus, they're able to hire the best accountants and lawyers to make sure they pay as little tax as legally possible. Gary Cohn, one-time chief economic advisor to Donald Trump, allegedly said, 'Only morons pay the estate tax.'[26]

Meanwhile, those who have to work for a living find their income relatively stagnant, offering little chance of building a legacy for their descendants. Thomas Piketty writes:

> 'When the rate of return on capital exceeds the rate of growth of output and income, as it did in the nineteenth century and seems quite likely to do again in the twenty-first, capitalism automatically generates arbitrary and unsustainable inequalities that radically undermine the meritocratic values on which democratic societies are based.'[27]

Financial crises (and other disasters) only exacerbate the problem. As Professor Mariana Mazzucato wrote in a 2020 article entitled 'Capitalism After the Pandemic':

> 'For too long, governments have socialized risks but privatized rewards: the public has paid the price for cleaning up messes, but the benefits of those cleanups have accrued largely to companies and their investors.'[28]

I'd like to repeat that I am not saying that capitalism is itself a bad thing. Its ascent has certainly correlated with (although not necessarily caused) a significant rise in living standards across the globe. When I first started writing on my blog about capitalism's pitfalls, I received emails from several friends whose income brackets may not have been in the 1 per cent, but were almost certainly in the top 5 per cent. Mention was made of 'Marxist tendencies'. (In all honesty, I don't know enough about Marx to say for sure whether I have anything in common with Karl. More likely Groucho.)

These concerned friends are good people, and I believe them when they say that in the course of their international travels over the decades they have seen massive improvements in living conditions in the countries they visited. But I also believe Jason Hickel when he says that the rise in income has not been fairly distributed, and that many people still lack adequate food, potable water, basic literacy and access to medicine.

Money Turns People into Jerks

A few years ago, I was on a weekend retreat with sixteen women. One of us, Alex, led an exercise that still gives me pause for thought. She handed out one playing card to each of us. We weren't to look at the card, but were to place it against our forehead with the number side facing outwards so others in the group could see it. We then had five minutes to interact with each other, and we had to behave according to the denomination of the other woman's card.

So if she was holding an ace, king or queen, we were to treat her as an important person worthy of respect and esteem. If she was holding a two or a three, we were to treat her as the lowest of the low. At the end of the five minutes, still without looking at our own card, we had to line up according to where we thought we lay in the social order, from high to low. With an uncanny degree of accuracy, we knew our place, and shuffled into almost exactly the correct sequence.

Here are some interesting things I noticed.

> The high-ups quickly started acting entitled, even though the cards had been handed out at random and they didn't know what their cards were. The low-downs were initially subservient, but it wasn't long before we (yes, I was a lowly three) were banding together and plotting to overthrow the kings and queens.
>
> We were exquisitely sensitive to our social ranking. Without knowing our own card, we knew exactly where we stood in the pecking order based on how others had treated us.
>
> It was almost scary how, regardless of our real life circumstances, and with no special acting abilities, we easily slipped into role as a high-up or a low-down.
>
> We were then given new cards and did the round again. This time I was a queen. It definitely felt better than being a three.

So how does this show up in real life, where some people are handed a high card at birth, while the vast majority are handed low cards?

There is ample research to back up the theory that rich people are more likely to act like entitled jerks, less likely to show empathy. As the American poet and satirist Dorothy Parker apparently said, 'If you want to know what God thinks of money, just look at the people he gave it to.'

Social psychologist Paul Piff's TED Talk, 'Does Money Make you Mean?', describes the studies done by him and his colleagues at the University of California, Berkeley. They brought more than 100 pairs of strangers into the lab and randomly assigned one of the two to be a rich player in a rigged game of Monopoly. They got twice as much money; when they passed Go, they collected twice the salary; and they got to roll both dice instead of one. The researchers retreated from the room to watch through hidden cameras what transpired.

Pretty quickly, the privileged player started acting in a more domineering way, helping themselves to the lion's share of the snacks, using assertive body language, smacking their playing piece down on the board with great aplomb, and being rude to their relatively disadvantaged fellow player. Most tellingly, when asked by the researchers afterwards about their experience of the game, the privileged player tended to dwell on their superior skill and strategy, conveniently forgetting that the game had been rigged from the start.

The same team ran a study using volunteers to stand at a crosswalk, waiting to cross the road. In California, where they ran the test, it's the law to stop for a pedestrian at a crossing. What they found was that as the expensiveness of a car increased, the drivers' tendency to break the law increased as well. None of the cars in the least expensive car category – the old bangers – broke the law. Close to 50 per cent of the cars in the most expensive vehicle category just kept right on driving, ignoring the pedestrian waiting to cross. BMW drivers were specifically mentioned.

Satisfying as it may be to confirm our prejudices about people who drive flashy cars, there are more serious issues going on here.

First, it tends to be those same privileged people who hold the most power in government. In the US, it's billionaires. In the UK, it's Old Etonians. If wealth is inversely proportional to empathy, what kind of a world are these rich and powerful people going to create?

Second, inequality understandably leads to unrest. Nick Hanauer, himself a rich guy, uses his TED Talk to issue the warning, 'Beware, Fellow Plutocrats, the Pitchforks are Coming', pointing out that massive inequalities have historically never been sustainable. Just ask Marie Antoinette.

Third, inequality fuels conspicuous consumption among the have-nots as well as the haves. Studies have shown that if you live in a more unequal area, you are more likely to spend money on an expensive car and shop for status goods.[29] Inequality drives up personal debt as people try to appear richer than they are. This is damaging to the environment and damaging to mental health. A *Guardian* article noted:

> 'Our rich society is full of people presenting happy smiling faces both in person and online, but when the Mental Health Foundation [in the UK] commissioned a large survey last year, it found that 74% of adults were so stressed they felt overwhelmed or unable to cope. Almost a third had had suicidal thoughts. 16% had self-harmed at some time in their lives ... And rather than getting better, the long-term trends in anxiety and mental illness are upwards.'[30]

This is what author John de Graaf has dubbed '**affluenza**, n. a painful, contagious, socially transmitted condition of overload, debt, anxiety, and waste resulting from the dogged pursuit of more'.[31]

So maybe the revolution isn't coming, after all. Depressed, overwhelmed, anxious people don't tend to reach for their pitchforks. They withdraw from society and self-medicate with alcohol, TV and video games, or indulge in retail therapy. Never before have there been so many ways to distract ourselves from harsh reality. And economic reality is getting harsher all the time.

The Inescapable Logic of the IPAT Equation

IPAT might sound cute, like 'I pat my head and rub my stomach', but I actually find it rather scary in its ruthless logic. It's an equation to describe the factors contributing to environmental impact, credited to biologist Paul Ehrlich:

Impact = **P**opulation × **A**ffluence × **T**echnology

I was curious to see how these figures had changed since my birth at the tail end of 1967.

Population: has risen from 3.5 billion in 1967 to 7.9 billion at the time of writing in 2022.

Affluence: globally the average GDP per capita has risen from $653 (current US$ equivalent) in 1967 to $10,926 in 2020.

Technology: is the rogue variable. It can drive down impact by increasing efficiency. But, before you get your hopes up, the Australian environmentalist Paul Gilding argues convincingly that technology has never in the history of humankind evolved fast enough to counterbalance the current growth in Population and Affluence.[32] We've had massive, *massive* advances in technology over the last fifty years, and yet our overall environmental impact is still increasing. This phenomenon is sufficiently well recognised to have a name: the Jevons paradox. According to this, increases in efficiency due to technological progress are more than outweighed by increasing demand due to lower costs, so the overall amount of the resource being used still goes up.

So, we can have sustainable impact, and we can maintain the current population, and we can live affluently, but we can't have all three. Pick two.

There are, of course, those ecomodernists who believe that technology *will* come to our rescue, that we really can have it all. My question to them would be … when? Our economic expansion is now butting up against, and in some cases overshooting, the carrying capacity of the Earth, as Kate Raworth illustrates so clearly in *Doughnut Economics*.

If technology is going to save the day, it had better get a move on and find a way to overcome Jevons. Personally, I would rather not gamble the future of humanity on imaginary, hypothetical and entirely unprecedented advances in technology. To those who propose that we technologise our way out of the unsustainable corner we've got ourselves into, good luck. I wish you well, and I hope you succeed. But I'm not banking on it.

I would rather put my time and energy into the human element of the equation. There are already a number of variations on IPAT, but I'd like to propose this one: IPATW (although obviously it's harder to pronounce than the original).

Impact = **P**opulation × **A**ffluence × **T**echnology × **W**isdom

If you believe, as I do, that the measure of a society is how it treats its neediest members, then their suffering reflects badly on us all. Wisdom could be the saving grace we're looking for. It's free of charge, it's renewable, it has zero environmental impact, and all of us have it (well, almost all).

As Tim Jackson put it when we spoke for my podcast:

> 'Those who benefit from inequality have no desire at all to do anything about it, and those who suffer from inequality have no power to do anything about it … What we're looking for is a politics brave enough to give voice to that muted outrage, to give it a place to go … We're still looking for those brave political leaders who are prepared to articulate that, and be the mouthpiece, to be the representatives of those who don't have the power to articulate, and don't have the voice to be heard.'

But where are those wise and brave political leaders? And could they ever prevail in our current political system?

Chapter 3

Politics

'Politics is the gentle art of getting votes from the poor and campaign funds from the rich, by promising to protect each from the other.'

Oscar Ameringer

Growing up as a child in 1970s Britain, I noticed that the government was something that people enjoyed grumbling about. Everything was always the government's fault, from strikes to power cuts to potholes in the roads to any and all shortcomings in the National Health Service. The only thing that wasn't the government's fault was that other subject that British people like to grumble about, the weather, although I'm sure some people blamed that on the government too.

Nonetheless, I had the overall sense that the government was there to look after us, and it may do that job well or it may do it badly (opinions varied), but either way, caring for its citizens was what government was created for, and what it was trying to do to the best of its abilities.

Ah, for the innocence of childhood.

The (corporate) penny started to drop when I watched Annie Leonard's *The Story of Stuff* soon after it came out in 2007. This short animation made me see the world in a new light that suddenly made a lot more sense than the old light. One line struck me particularly hard:

'The reason the corporation looks bigger than the government is that the corporation *is* bigger than the government. Of the 100 largest economies on earth now, 51 are corporations. As the corporations have grown in size and power, we've seen a little change in the government, where they're a little more concerned in making sure everything is working out for those guys than for us.'[1]

The truth may be more blatant in some countries than in others – and I may even have thought at the time, 'Oh, poor old US, but that could *never* happen in the UK' – but the more I thought about this idea and tried it on for size, the more accurate a representation it appeared to be. Essentially, *The Story of Stuff* introduced me to the idea that we might think we live in a democracy, where delegates chosen by fair process represent the interests of their constituents in a central body equally beholden to fair process, but we would be mistaken. We live in a corporatocracy, where money talks, and vested interests have ways and means of getting their way.

But I'm getting ahead of myself. I'll start off with some of the less controversial aspects of politics, and work back up to corporatocracy.

In their book *Crowdocracy*, Alan Watkins and Iman Stratenus set out the most fundamental shortcomings of the current model of democracy as follows.

When is a Democracy not a Democracy?

Mark Twain said, 'If voting made any difference, they wouldn't let us do it.' His view may be cynical, but it's hard to disagree. It's been said that we get the politicians we deserve, but often we don't even get the ones we voted for.

The discrepancy between who we vote for and who we get is largely a result of gerrymandering, the practice of manipulating the boundaries of electoral districts to favour the party in power, combined with the first-past-the-post system, which means that the number of seats a party wins bears little relation to the number of votes it received. In 2015, the British Conservative government won the election with only 37 per cent of the vote – the rest of the electorate voted for a different party or didn't vote at all. The Greens and UK Independence Party got more than 5 million votes (out of an electorate of 46 million), but got only one seat apiece. Meanwhile, in Scotland, the Scottish National Party received 50 per cent of the votes but won 95 per cent of the seats.[2]

To look at it a different way: in the UK General Election in 2019, the Conservatives won a seat for every 38,264 votes, while a Labour candidate needed 50,837, a Lib Dem 336,038 and a Green 866,435. The only party with an easier ride than the Tories was the Scottish National Party, which won a seat for every 25,883 votes.[3]

On the other side of the pond, in different years Al Gore and Hillary Clinton polled 51 per cent and 56 per cent respectively, but still lost in presidential elections. So it's clear that we're a long way from representative democracy in the UK and the US. And I haven't yet mentioned the role of social media and big data, as exposed in the Cambridge Analytica case where millions of Facebook users' data was collected without their knowledge or consent to use for political advertising. Cambridge Analytica may have gone away, but I'm sure their strategies and sharp practices haven't.

It gets worse. We in the UK might like to think that corruption is something that happens elsewhere, and imperfect as the democratic system may be, policies generally align with the public interest. We would be wrong. Strings get pulled – and pulled energetically – in UK policy making; for example, by property developers who would prefer not to be inconvenienced by obligations to preserve natural habitats for biodiversity.

In the US, a study looked at more than twenty years of data and found that, when it came to the richest 10 per cent of Americans, if fewer than 20 per cent of the elite supported a policy change, it happened only about 18 per cent of the time. But if 80 per cent or more of them were in support, the change happened 45 per cent of the time. Outside of the elite, it was a different story: 'The preferences of the average American appear to have only a minuscule, near-zero, statistically non-significant impact upon public policy.'[4]

Small wonder that many people feel their government doesn't represent their views and interests. Chances are, they're right.

Powerful Vested Interests

In the US, there is a strong correlation between a candidate's budget and their chances of electoral victory. In the Senate in 2014, 91 per cent of the candidates who spent the most money won. Donald Trump in 2016 was the first president since JFK in 1960 to win the presidential election despite a smaller budget.[5]

So fundraising is a large part of the candidate's job. Ronald Reagan wryly remarked, 'I used to say that politics was the second-oldest profession. I have come to know that it bears a gross similarity to the first.'[6] Do we really believe that donors expect no favours in exchange for their generosity, or that a politician is immune to the hope of future largesse?

The sums of money involved are vast. In the five years to 2015, the 200 most politically active companies in the US spent $5.8 billion influencing the government, and received $4.4 trillion in taxpayer support[7] – so their money gave an excellent return on investment. In fact, they got a better ROI on lobbying than they did on research and development,[8] so where are they more likely to invest their money in the future? It seems obvious.

In the UK, the lobbying industry is estimated to be worth £2 billion. The late Nye Bevan MP referred to 'gastronomic pimping', in which MPs are handsomely wined and dined. As with a romantic dinner, the aim of the game is seduction. The former prime minister David Cameron named it, even if he didn't do anything to remedy it, saying, 'It arouses people's worst fears and suspicions about how our political system works, with money buying power, power fishing for money and a cosy club at the top making decisions in their own interest.'[9] Jimmy Carter was even more blunt, describing American presidential politics as 'an oligarchy with unlimited political bribery'.[10]

Worst of all, the dynamic becomes self-reinforcing. As the rich exert political influence, market regulation and top tax rates decrease, so they become even richer and are able to buy even more influence.

As well as direct lobbying, the rich are able to finance academic studies and think tanks that cast doubt on sound science and sow seeds of confusion in the collective public mind. If you ever see reports of a 'study' claiming something a little surprising, check who funded it, and then who funded them. Follow the money. The research may not be as objective as it might at first appear.

When the rich are also able to control the mass media networks, this problem becomes even more entrenched. Not only are they able to influence politicians, but they are able to influence what voters think of politicians, and hence drastically improve the chances of their favoured candidates. In the last forty years, in the US, fifty media corporations have been consolidated into five, concentrating power in a small number of hands. In the UK in 2021, just three companies (News UK, Daily Mail Group and Reach) dominated 90 per cent of the national newspaper market.[11] As the American human rights activist Malcolm X warned, 'If you're not careful, the newspapers will have you hating the people who are being oppressed, and loving the people who are doing the oppressing.'[12]

This pattern of consolidation through acquisition has also been replicated in the fields of technology, pharmaceuticals, banking, oil and consumer goods. All of these are cause for concern, but if knowledge is power, then having control over the public dissemination of knowledge is the ultimate power.

Democracy Fosters Superficial and Inadequate Focus on the Issues

The night that Barack Obama was inaugurated, a group of high-powered Republicans met and agreed to 'show a united and unyielding opposition to the president's economic policies'.[13] The increasingly adversarial nature of politics in both the US and the UK (and many more countries besides) makes it harder and harder to find the optimal outcome for the country, when the major parties refuse to agree with each other on anything at all.

Too often mature political debate on important issues devolves into cheap point-scoring – witness the schoolyard jeers of Prime Minister's Questions in the UK Parliament. There have been occasional moments of witty repartee, such as when Benjamin Disraeli (Conservative leader) was asked to explain the difference between a misfortune and a calamity, he shot back the answer: 'If Gladstone [Whig leader] fell into the Thames, it would be a misfortune. If anybody pulled him out, that, I suppose, would be a calamity.'

But on balance, I'd gladly give up the adolescent banter in preference of a proper, grown-up conversation in a sincere quest for the optimum way forward on matters of national importance.

Democracy Fosters Division

A great many politicians would appear to operate from the principle that 'my enemy's enemy is my friend', so it is in their interests to identify a common foe in order to unite their voters. In 1982, the Argentinians did Margaret Thatcher a great favour when they invaded the Falkland Islands, a small huddle of wind-blasted rocks that was home to 1,847 people, 8,000 miles away in the South Atlantic – not somewhere off the north coast of Scotland, as most Brits would have guessed at the time, so little did they know or care about the Falklands before the Argentinian occupation.

The resulting war gave the Iron Lady an opportunity to showcase her best assets – single-minded decisiveness and determination. She would stop at nothing to rescue the islanders and their 400,000 sheep from the evil Argies, as the invading Argentinians became known in the popular press. It was a major public relations coup, and saved her political career.

The same strategy continues to reap rewards; the Scottish National Party benefits from shared – and historically understandable – animosity towards the English, Brexiteers benefited by stoking the fires of xenophobia, Donald Trump united many Republicans behind the rallying cry of 'lock her up', and so on. This may achieve the short-term goal of gaining power, but it is power over a country tragically (and let's hope not irredeemably) divided.

Politicians are Struggling Under the Weight of Escalating Complexity

Politicians are human beings (allegedly), and the world is complex, interconnected and not easily divisible into tidy administrative chunks. It's also rapidly evolving, with no issue standing still for long enough for governments to get fully on top of the situation – especially when a massive wrecking ball like Brexit comes along. No matter how smart our politicians are (see following point below), it is virtually impossible for any one person to wrap their head around their department's remit in the relatively short time they have to get up to speed. So spare a compassionate thought for those in government. It's not an easy job.

But if our world has now become so complex that a minister can't be expected to fully understand his or her brief, that has serious implications. First, every decision has unforeseen consequences, and the less the grasp of the situation, the greater the chance of adverse impacts. Second, it makes the minister very reliant on their department's civil servants, who weren't elected, and who may have their own agendas.

Democracy is not a Meritocracy

Tony Blair appeared for an onstage interview at Yale while I was there in 2012. After pointing out that before becoming prime minister in 1997 he had held not so much as a Cabinet position, he said, 'On the day you're voted into power, you're at your most popular and your least competent. On the day you're removed from power, you're at your least popular and your most competent.' Cabinet appointments, in the UK at least, are based less on relevant expertise than on whether one's star is rising (Foreign Secretary) or falling (Culture). And sometimes a minister has just got the hang of, say, Transport, when they get whisked to the Exchequer, and have to master their brief from day one with zero tolerance for failure.

Not that voters vote according to their perception of someone's intelligence or expertise. Voters are more likely to be swayed by someone's charisma, ubiquity or media image than by their ability, character and integrity. Televised debates reward snackable soundbites rather than actual knowledge or policy insights. Albert Mehrabian's oft-cited UCLA study[14] reported that gestures count for a 55 per cent of the impact a speaker has on an audience, while tone of voice makes up 38 per cent. Actual words count for only 7 per cent. Add that up, and the non-verbal part of a presentation accounts for a staggering 93 per cent of its impact.

Princeton psychologist Alex Todorov established that the impression formed of a political candidate *within the first second* predicts 70 per cent of US Senate and gubernatorial race outcomes.[15] Psychologists tell us that we make decisions primarily with our emotions, using our brains to create an after-the-fact justification for deciding the way we did. So a politician's haircut, clothes, level of self-confidence or regional accent (or lack thereof) are more likely to influence our vote than the actual words that come out of their mouth.

Or, increasingly, the electorate is largely so busy and distracted that they are likely to vote in accordance with how they have always voted, or how their friends are voting, rather than taking time to do due diligence on the candidate's view on important issues and policies.

Politicians Don't Have a Voice Either – Their Vote is 'Whipped'

For non-UK readers, the 'whip' isn't actually a physical whip (although it may sometimes have the same effect). The term refers, in fact, to a political official appointed by their party to ensure their members of parliament toe the party line in a vote. And it's not just party loyalty that tugs on a politician – they have many other loyalties to take into account.

In fact, when you look at all the conflicting pressures on an elected politician, we might almost start to feel some sympathy for them. It seems that whichever way they turn, they are bound to lose the support of somebody – their constituents, their party, their peers, their donors, or maybe even their own family and friends.

Not only are they beholden to the relationships within which they operate, they are also creatures of their times. The Overton window is the name given to the limited range of policies that is politically acceptable to the mainstream population at a given time. Being too far ahead or behind the wave of current opinion can make the difference between a glittering career and a tarnished one. And a politician who thrives in wartime may fall out of favour in peacetime (witness Winston Churchill), or simply pass their use-by date, as Margaret Thatcher discovered when the single-mindedness that had been her trademark became her fatal flaw.

Of course, regardless of their competency or lack thereof, they will get blamed for everything. Barack Obama said, 'As president, you're held responsible for everything, but you don't always have control of everything.'

Self-interest and Scandal

In his *Devil's Dictionary*, Ambrose Bierce caustically defines politics as 'A strife of interests masquerading as a contest of principles. The conduct of public affairs for private advantage.' Possibly because politicians generally get paid less than their peers in banking or industry, they often feel entitled to a few perks on the side of varying degrees of legitimacy.

Since leaving office in 2007, Tony Blair has accumulated an estimated £60 million in earnings,[16] while Bill and Hillary Clinton made $240 million between 2000 and 2015.[17] Vladimir Putin is estimated to be worth

$200 *billion*,[18] which would make him among the top few richest people on Earth – and we can only speculate as to how he came by his fortune.

Arguably, they are entitled to their wealth – politics is a brutal business, and the CEO of Time Warner makes up to $50 million a year[19] for running a Fortune 100 company, which must surely be a less demanding job than running a country. But the eager acceptance of lucrative roles in the very industries they were formerly regulating inevitably leads to speculation as to whether their motivation was service to their country or service to themselves.

Better Politicians? Or a Better Politics?

One way and another, democracy has issues. There are, of course, many and varied forms of democracy, some of which appear relatively effective – Switzerland, Singapore and the Nordic countries seem to have well-functioning models. But for many of us, the world of politics seems distant, dysfunctional, rife with egos and self-serving if not downright corrupt. What do we want – better politicians? Or a better politics?

Both, of course. But which is the chicken and which is the egg? As the comedian Billy Connolly quips, 'The desire to be a politician should bar you for life from ever becoming one.' How do we create the kind of politics that will attract principled individuals who genuinely want to be of service to the greater good?

Two American women, Barbara Marx Hubbard and Marianne Williamson, both primarily known as spiritual teachers, were brave enough to wade into the murky waters of the political arena. Barbara Marx Hubbard's 1984 run for the US vice presidency wasn't serious – she simply wanted the opportunity to address the Democratic National Convention to invite them to raise their game and dare to dream of a better world. Her obituary in *The New York Times* led with her political bid, stating that she wished to escape the pervasive negativity of politics in favour of a positive vision of the future of humanity. 'We must combine our compassion with our creativity,' she said. 'We must initiate a new process in democracy to identify our positive options, discover our potentials and commit our political will to long-range goals.'[20]

This vision may sound appealing, but both she and Marianne Williamson were widely ignored, dismissed, and particularly in Williamson's case, derided as New Age dreamers. An article in *The Guardian* referred to Williamson as a 'celebrity self-help guru accused of being a "dangerous wacko" who just wants the United States to "harness love"',[21] running a 'bizarre and mesmerizing campaign'.[22]

On the one hand, we seem to wish for a system that supports candidates of integrity and good character. We admire fictional political characters, like Jed Bartlet of *The West Wing*, for taking a stand for principle over pragmatism. The very fact that we are so often disappointed by our politicians implies that, despite everything, we still have high expectations of them. We want to look up to them as role models, as protectors, as the grown-ups in the room who will ensure our world is orderly, secure and fair. Despite our carping, complaining and political satire, politicians still dominate Gallup's charts of most admired men and women.[23]

But, on the other hand, we as voters may have a story about what kind of person can succeed as a politician, a story that is heavily influenced by perceptions derived from the media, not only in the news, but also in comedy. Have we internalised a story that a politician needs to have a big ego, ambition, ruthlessness and finely-honed powers of political manoeuvring? Do we fail to take seriously candidates who offer a gentler, more inclusive way? Do we dismiss them as too idealistic to make it in the cut-throat world of politics?

There are, of course, many in public office who are remarkable for their dedication, hard work and sense of fairness. But, in general, we seem to have an inescapable conundrum. When discussing economics, we saw how those with the power to change the system have no interest in doing so, because it is the very same system that conferred their power. In politics, we see a parallel dynamic that often appears so hostile, corrupt and toxic that only those with the hide of a rhino or the ambition of Vlad the Impaler can survive, while the people who hold paramount the values of compassion, equality and long-term thriving we expect to be chewed up and spat out by the current political machine. It often seems that politics is a self-perpetuating system. The names might change (although in the case of Bushes and Clintons, not so much) but the power structures remain the same.

Although it can be fun to laugh at our politicians, to roll our collective eyes at their venality, to react with outrage to their latest lies and shenanigans, on balance, I'd prefer to be able to return to my childish idealism. I'd like to be able to admire the people we pay to serve us, and have faith that they are good people with honest intentions. Isn't that what democracy is supposed to be?

The fundamental problem is that these power structures are based on a win–lose dynamic, *power over* rather than *power with*. It seems that if we had wanted to design our political, economic and social structures in such a way to guarantee rampant inequality and environmental destruction based on short-term thinking, we have succeeded magnificently. They served their purpose for a while, but now the goals, processes and culture of the old paradigm are past due for an overhaul. They are founded upon the principles of domination and disconnection, which lie at the root of all our troubles.

Chapter 4

Power

'We are living now inside the imagination of people who thought economic disparity and environmental destruction were acceptable costs for their power.'[1]

adrienne maree brown

It was my second visit to Riane Eisler's home, a sprawling, single-storey house in a quiet suburban neighbourhood of Carmel-by-the-Sea on California's Monterey Peninsula. It was a Sunday in the summer of 2019 when I rang the doorbell, table-top tripod in hand, ready to film our conversation. So I was a little taken aback when she opened the door looking very different from the well-groomed, immaculately coiffed Riane I'd met the previous year. She was wearing a kind of bathrobe or housecoat, and a surprised facial expression.

It was totally my fault. I'd had to switch my plans around at short notice and could have sworn that I'd confirmed this with her by email, but a later search of my sent messages proved I would have sworn wrong. Once she had recovered her composure, and I had apologised profusely, she was gracious enough to invite me in and I waited in her sunny, book-filled sitting room while she made herself camera-ready. A true professional, she gave no hint of any irritation at me barging into her peaceful Sunday afternoon.[2]

Riane Eisler was born in Austria in 1931. The growing power of the Nazi Party and escalating hostility against Jews reached a crescendo on Kristallnacht in 1938, when 7,000 Jewish businesses across Germany and Austria were damaged or destroyed. After Riane's father was pushed down the stairs by the Gestapo that night, he took his family and fled to Cuba on one of the last ships that managed to leave Europe before war broke out.

Later on, now living in the US and reflecting on subsequent events in her home country, Riane wondered how the Nazis could have overpowered people's natural humanity, sweeping them up as active participants in an abhorrent regime. She set out to answer some searing questions:

> 'Why do we hunt and persecute each other? Why is our world so full of man's infamous inhumanity to man – and to woman? How can human beings be so brutal to their own kind? What is it that chronically tilts us toward cruelty rather than kindness, toward war rather than peace, toward destruction rather than actualisation?'[3]

Her conclusions, documented in her 1987 book *The Chalice and the Blade*, was that the early twentieth-century culture of Austria and Germany was patriarchal and authoritarian, and this pattern of one person or group having power over another was instilled in children from the moment they were born, with the father as the head of the household and the mother in a correspondingly subservient role. Children simply grew up accepting the idea that men were superior to women and children, and were therefore entitled to command and control them. When this is the blueprint of a society, with one group positioned above another, then it is entirely consistent with that blueprint to believe that one race is superior to another, one religion superior to another, and one country superior to another. Dominating and being dominated is understood to be the natural order of things.

> 'Once the function of male violence against women is perceived, it is not hard to see how men who are taught they must dominate the half of humanity that is not as physically strong as they are will also think it their "manly" duty to conquer weaker men and nations.'[4]

Gender-based discrimination is just one of countless ways that humans divide ourselves, one group from another, and then place one group in a more privileged position. These divisions are almost always based on characteristics such as ethnicity, religion, gender, sexual orientation or nationality – in other words, inherited or otherwise innate characteristics, rather than anything we chose to be or do.

I first read *The Chalice and the Blade* in 2018, over thirty years after it was first published, and it instantly became one of several books that fundamentally influenced the way I see the world. It confirmed something that had been tickling around the periphery of my perception, but hadn't yet been consciously articulated.

The book corroborated and expanded on a hunch I'd long had, that gender equality and sustainability are deeply intertwined. I'd thought that maybe this was because the stereotypically feminine qualities of compassion and nurturing would lead to greater consideration for the non-human and future human inhabitants of planet Earth. But Riane's perspective goes deeper. She distils her findings into her Cultural Transformation Theory, according to which there are two alternative models of society:

1. Dominator model: ranking of one half of humanity over the other, in either patriarchy or matriarchy
2. Partnership model: social relations based on linking rather than ranking, so groups can be different but equal; neither inferior or superior to each other

After a lengthy examination of the archaeological evidence,[5] she concludes that the original direction of our cultural evolution was towards partnership, but following a period of chaos and disruption around 5,000 years ago there was a fundamental shift towards domination, and we moved from a life-generating and nurturing view of the universe (chalice/grail) to a view that venerates the power to take life (the blade). The sexual imagery of the chalice and the blade is obvious. I'm sure you can think of many examples of the dominator mindset alive in the world today, without too great a stretch of the imagination.

The Chalice and the Blade was a revelation, one of those special books that helps a nonsensical world make more sense. I get really curious about the truth that hides behind superficial appearances, like an annoying child who keeps asking, 'But why? But why?' until she has dug her way all the way down to the ultimate why (or until the exasperated parent says, 'Just *because*!'). Applying the dominator lens to the relationship between humans and nature seemed get to the root of what ails us.

Domination paradigms don't work – not for the dominated, obviously, but also not for the dominating. As we saw in relation to economic inequality, having the upper hand comes with its own pressures, not least that the only way is down. The essential point is that a system that doesn't work for everybody, doesn't work for anybody.

So how did we end up in this power structure of the dominating and the dominated? How does it show up in our world? And what are the implications?

A History

The narrative of domination started early in human mythology, with the Genesis creation story. According to Genesis 2, Eve was created from Adam's hand-me-down rib, to give him assistance, company, and presumably sex. So Eve was essentially created to serve Adam's needs. Or, according to some versions of the creation myth, Eve was actually the second wife of Adam, with Lilith being the first. Lilith was made by God out of dust, rather than cast-off bones, and according to one version of the story, things went awry between the first couple when Adam tried to exercise dominance over Lilith. They were about to make love and he told her to lie down so he could mount her. She refused. He became enraged, and she fled the Garden of Eden. Long story short, her punishment for asserting her equality was to have her children killed by angels, and to be forever blamed for stillbirths, cot deaths and wet dreams.[6] The misogynistic moral of the story is ominously clear.

Eve didn't fare much better. According to Sharon Blackie, mythologist and author of *If Women Rose Rooted*:

> 'In this story [Genesis], the first woman was the cause of all humanity's sufferings: she brought death to the world, not life. She had the audacity to talk to a serpent. Wanting the knowledge and wisdom which had been denied her by a jealous father-god, she dared to eat the fruit of a tree. Even worse, she shared the fruit of knowledge and wisdom with her man. So that angry and implacable god cast her and her male companion out of paradise, and decreed that women should be subordinate to men for ever afterwards.'[7]

Blackie says that over time this story made her angry, because it set the tone for 'a world in which men still have almost all the real power over the cultural narrative – the stories we tell ourselves about the world, about who and what we are, where we came from and where we're going – as well as the way we behave as a result of it'.[8]

If this doesn't ring true for you, I recommend you take a look at *Invisible Women* by Caroline Criado-Perez. It's an eye-opening – and occasionally eye-popping – insight into the many ways that our world is still designed for a default human who happens to be male, and the way this impacts on everything from clinical trials to urban snow-clearing to the design of police body armour, in ways that directly affect women's physical health and safety as well as their social standing. Criado-Perez launched a campaign to have Jane Austen put on English £10 banknotes in an attempt to redress the balance, the only other human with a vagina on banknotes being, of course, the late Queen Elizabeth II. (I can't believe I just put 'vagina' and 'the Queen' in the same sentence.) Her crusade provoked abusive messages threatening her with rape and murder, which surely only underscored her point that women are still a long way from equal status.

I want to emphasise that, guys, this isn't having a go at you. This isn't to blame the current generation of men for the situation we're in – many men are wonderful champions and allies of women's rights. History is as it is. And for the record, Sharon Blackie is not anti-men. She has even married a couple of them (not at the same time, obviously). But she is anti-patriarchy, meaning she is frustrated by power structures that supposedly advantage men at the expense of women (although, paradoxically, patriarchy can take a heavy toll on men as well, burdening them with stereotypical images of what a man is – or is not).

The Genesis gender story just happens to be the first example of one group being arbitrarily assigned a higher status than another. Actually, that's not true – allow me to correct myself. Before Genesis gets as far as serpents and figleaves, it has already given humans permission to dominate nature:

> 'Let us make man in our image, after our likeness: and let them have dominion over the fish of the sea, and over the fowl of the air, and over the cattle, and over all the earth, and over every creeping thing that creepeth upon the earth.'[9]

Some biblical scholars challenge this translation, asserting that the relationship between humans and Earth was meant to be less like dominion, more like stewardship.[10] Certainly, the *dominion over* version isn't working out so well for humanity right now, and *partnership with* would seem to be a better idea to secure our survival on this planet.

Ultimately, the gender narrative and the nature narrative are from the same root, and are equally dangerous and short-sighted. Sharon Blackie draws the parallel between gender domination and nature domination:

> 'The same kinds of acts that are perpetrated against us, against our daughters and our mothers, are perpetrated against the planet: the Earth which gives us life; the Earth with which women have for so long been identified. Our patriarchal, warmongering, growth-and-domination-based culture has caused runaway climate change, the mass extinction of species, and the ongoing destruction of wild and natural landscapes in the unstoppable pursuit of progress.'[11]

Fracking seems to me the consummate illustration of our rapacious dominion over Mother Earth – forced penetration followed by injection of unwanted fluids. It makes me shudder.

Where else does this narrative of domination show up? Just about everywhere. It has been the meta-narrative of human civilisation for many millennia. Let's first look at the three areas I covered in the opening chapters – environment, economics and politics – through this lens, and then get curious about how dominator strategies show up elsewhere. These subsequent section headings are inspired by Barbara Marx Hubbard's Wheel of Co-Creation[12] and represent the main facets of modern human society: our relationship with the world around us, our monetary value system, the way in which we govern ourselves, how we facilitate learning and the transfer of knowledge, the rules we choose to abide by and how we mete out 'punishment', and how we interact with and interpret the world through technology, science and religion.

Environment

I had an 'a-ha' moment at a site visit during a permaculture design course. We were at a private 2-acre permaculture garden, which to a casual observer looked less like a garden and more like wild land with a polytunnel on it – but that, really, is the point of permaculture, which is not about trying to control nature, but about working in harmony with it, even emulating it.

I was bemoaning the twin banes of my gardening life: bindweed and slugs. 'When you find yourself getting frustrated,' said Alison, the permaculturist, 'it's not nature that needs fixing, it's your attitude.' She went on to explain that everything has its role in the ecosystem, and when we go to war with any single species, we disrupt the system in ways we can barely begin to imagine. We try to bludgeon nature into submission with an arsenal of sprays, pellets, oil-based fertilisers and other weapons of biomass destruction. We might think it doesn't matter if we eradicate a species from the tiny fraction of the world that constitutes our garden, but the current ecological crisis is this attitude writ large, at planetary scale.

It's easy to see how this bellicose attitude came about. Nature can appear unfriendly, even dangerous – and I speak from personal experience. Early humans understandably wanted to protect themselves from famine and drought and extremes of hot and cold, so they systematically set about creating a more comfortable, secure life through agriculture and homes, and later supermarkets and supply chains. In the course of this determined attempt to conquer nature, most of us have become tragically disconnected from the natural systems that support life, and disconnection leads to disrespect and destruction. The novel *Ishmael* points out how humanity's aggressive food production methods have resulted in a holocaust:

> 'Kill off everything you can't eat. Kill off anything that eats what you eat. Kill off anything that doesn't feed what you eat ... everything in the world except your food and the food of your food becomes an enemy to be exterminated.'[13]

The original reason for agriculture was to prevent hunger, which is an admirable goal. Unfortunately it hasn't worked, at least for large proportions of the human population. One in nine people are hungry or undernourished, and 2.37 billion people did not have access to sufficient safe and nutritious food in 2020.[14] In 2021, 42 million people were on the brink of starvation. A multibillionaire is welcome to throw his money at the problem, but that won't fix the system that created it.[15]

Even when we're not actively killing nature, much of human activity is oriented around exploiting it, controlling it or dominating it. So thoroughly have we subjugated nature, we have forgotten we are a part of it.

Economics

As we've already seen, the current model of wealth distribution has led to unprecedented inequality, and according to the UN's *World Social*

Report 2020, the situation is still worsening for 70 per cent of the world's population.[16] Yet we have often been told, in the UK and US at least, that we live in a meritocracy, where anybody can rise to the top if they work hard enough. So if income is supposed to be a fair reflection of effort and intelligence, and a CEO is paid 350 times as much as a typical worker,[17] then are we to assume that the CEO has worked 350 times as hard, or is 350 times as smart? It seems unlikely. It's hard not to agree with Michael Sandel when he says, 'The dark side of the meritocratic ideal is embedded in its most alluring promise, the promise of mastery and self-making. This promise comes with a burden that is difficult to bear.'[18]

In his 2009 TED Talk, philosopher Alain de Botton puts this into historical context:

> 'In the Middle Ages, in England, when you met a very poor person, that person would be described as an "unfortunate" – literally, somebody who had not been blessed by fortune, an unfortunate. Nowadays, particularly in the United States, if you meet someone at the bottom of society, they may unkindly be described as a "loser". There's a real difference between an unfortunate and a loser, and that shows four hundred years of evolution in society and our belief in who is responsible for our lives. It's no longer the gods, it's us. We're in the driving seat. That's exhilarating if you're doing well, and very crushing if you're not.'[19]

He goes on to say that there are more suicides in developed, individualistic countries than in more partnership-oriented societies. And it's not just the poor that opt out of this cruel world. It's also those who by many standards would seem to be the haves, rather than have-nots, who take their own lives. Celebrities and CEOs may have different problems from those further down the ladder of perceived success, but they still have them.

For as long as there is inequality and poverty in the world, insecurity looms for all, even the apparent winners in the game of life. There is a new word that has emerged in the last twenty or thirty years: the precariat, meaning those whose employment and income are insecure. But maybe the most precarious position of all is to be one of those balancing atop the tottering pyramid of the precariat. The higher they are, the further they fall when the pyramid crumbles.

In the current economic model, money is distributed by several means, all of which are unfair to a greater or lesser degree. Money might be inherited –

you are lucky enough to get born into a wealthy family and acquire wealth when your parents die. Or you sell your time for money, but some people's time is deemed more valuable than others', and this valuation is heavily influenced by factors such as a good education, verbal and numerical-logical skills, physical height,[20] and the ability to project confidence – not coincidentally the same traits that are more prevalent among the already wealthy. Or you create your own business, which is a risky endeavour, with more than half of start-ups going bust within the first few years. These are all very haphazard ways of distributing wealth, and cause a great deal of stress, envy, insecurity and unhappiness.

If the prevailing ethos is 'who dies with the most toys, wins', then even he or she who has more toys than they could ever play with in one lifetime doesn't win if somebody else has more. Whether it's a country's GDP or an individual's net worth, money has become the dominant metric by which we decide who is top dog, with happiness, fulfilment and other intangibles relegated to the margins.

Politics

Do we see a dominator paradigm in politics? Do I even need to answer this question?

In a democratic political system, the power is vested in the party that won the most recent election, and they then get to determine policy. In a communist political system, the power is vested in the party that has declared itself the only party with the authority to determine policy. Either way, once a party is *in power* (note the terminology), it mostly gets to do what it wishes without further consultation.

Some systems are more collaborative than others, running referenda from time to time, but the basic model is still to have a clear line between the government and the governed. You may say that no other option is workable, given the size of populations and the complexity of modern civilisation. But is this really true? Given modern communications technology, is there a way that citizens could have a greater say in determining the policies that affect their lives? More on this later.

Not only do political parties seek to have power over their country, they also seek power over their opponents – initially in the election and thereafter while in office. In the US and the UK in the mid-twentieth century, cross-party cooperation was still possible, but a 2015 study of US House of Representatives from 1949 to 2012 found that 'despite short-term fluctuations, partisanship or non-cooperation in the U.S. Congress has been increasing exponentially for over 60 years with no sign of abating or reversing'.[21] Cheap shots and power plays serve nobody, least of all the electorate that (mis)placed its faith in the democratic process.

Education

What is education meant to achieve? Is it meant to help students to reach their highest potential? The cynic could be forgiven for thinking it is better designed to teach them to conform to authority and become obedient little consumers.

The dominant education systems, at least in the US and most of Europe, quickly decide who is top and who is bottom based on a narrow range of skillsets, even though a well-functioning community needs a broad range of capacities. Large class sizes and endless exams result in teaching to the test, rather than encouraging curiosity, creativity and critical thinking.

According to Harvard psychologist Howard Gardner, there are at least eight types of intelligence: visual-spatial (e.g. engineers, architects, mechanics), linguistic-verbal (e.g. writers, translators, anybody who has to write reports), interpersonal (e.g. caring professions), intrapersonal (e.g. philosophers, therapists), logical-mathematical (e.g. actuaries, traders, accountants), musical (no explanation needed), bodily-kinaesthetic (e.g. dancers, sports people, artists and sculptors) and naturalistic (e.g. conservationists, gardeners, farmers).

But overwhelmingly, our traditional education system recognises and rewards primarily just two: linguistic-verbal and logical-mathematical. We'd be scuppered without our plumbers, electricians and car mechanics, and we'd be impoverished without our artists and musicians, but practical skills and capacities are often rated – and paid – lower than more cerebral pursuits. Not coincidentally, the intelligences that are valued over others correlate strongly to those that used to be gentlemen's pastimes, rather than those of the merchant or working classes.

Sir Ken Robinson, in the most widely watched TED Talk to date, says:

> 'Academic ability ... has really come to dominate our view of intelligence, because the universities design the system in their image. If you think of it, the whole system of public education around the world is a protracted process of university entrance. And the consequence is that many highly talented, brilliant, creative people think they're not, because

> the thing they were good at at school wasn't valued, or was actually stigmatized. And I think we can't afford to go on that way.'[22]

My cynical suggestion that education is as much about engendering conformity as it is about imparting information intends no disrespect to teachers, who all too often have a highly challenging and inadequately paid job. But it has been shown that children can be curious and highly motivated to learn with minimal supervision.

Sugata Mitra won the TED Prize for his talk on Self-Organised Learning Environments. In early 1999, he and some colleagues embedded a computer into a wall near their office in New Delhi. The area was a slum, inhabited by desperately poor people struggling to survive. The screen was visible from the street, and the computer was available to anyone who passed by. It had online access and a number of programs, but no instructions. Almost immediately, the slum kids showed up and started to play. Within six months they had learned how to operate the mouse, could open and close programs, and were going online to download games, music and videos. Mitra and his colleagues asked them how they had learned these sophisticated operations, and they replied that they had taught themselves.

Over the following ten years, Mitra and his team continued their research in different places and contexts. Everywhere, they found that children had enormous capacity to teach themselves and each other through random experimentation, teamwork and knowledge-sharing. Mitra observes:

> 'When working in groups, children do not need to be "taught" how to use computers. They can teach themselves. Their ability to do so seems to be independent of educational background, literacy level, social or economic status, ethnicity and place of origin, gender, geographic location (i.e., city, town or village), or intelligence.'[23]

His proposal is to allow children their freedom to learn, gently guided and corralled by 'grandmothers', whose main role is to offer encouragement and praise, rather than any specific expertise.

Of course, when schools began hundreds of years ago, there were no computers and few books, so the teacher was usually the only source of information other than the students' own experiences. But that is no longer true. As Margaret Mead said, 'Children must be taught *how* to think, not *what* to think.'

Justice

In Cathy O'Neil's book about the perils of algorithms, *Weapons of Math Destruction*, she offers a stark choice between two ways to use mathematical models driven by big data. These models enable authorities to identify the neighbourhoods and demographics facing the greatest challenges from crime, poverty and substandard education – do we then use that information to persecute and punish those people? Or to offer them the resources they so desperately need? It is up to society to choose which objective the data should serve.[24]

Too often, it seems that the objective we choose for the criminal justice system is domination and control, rather than support. The system is based on othering and judgement. The fundamental attribution error, a well-known cognitive fallacy, says that we tend to attribute another's actions (especially the actions we don't approve of) to their character or personality, whereas when we do them, we excuse our behaviour as being due to circumstances outside our control. In other words, we cut ourselves a break while holding others 100 per cent responsible for their actions. The justice system, from the police to courts to prison officers, is largely based on fundamental attribution.

The main theories of punishment are:

Retribution: an eye for an eye

Deterrent: to stop the criminal doing it again, and to make an example of them to others

Prevention: the criminal can't do it again while they're under lock and key

Compensation: the criminal wronged society, so their sentence is society's way to get its own back

Reformation: punishment will make the criminal a better person

You'll notice that none of these theories addresses the systems, situations or circumstances that led the criminal to commit the crime in the first place. Judgement of any sort, including judgements conducted in the name of justice, imply that *I am right and you are wrong*. Capital punishment for murder is the epitome of this: *you killed somebody, which is wrong, so we are going to kill you, but we are right*. Clearly, there are valid and necessary reasons why antisocial behaviour needs to be addressed, but can we be so sure that, if we walked a lifetime in the shoes of the criminal, we wouldn't have acted the same way they did?

Technology

Tristan Harris, former design ethicist at Google and now the president of the Centre for Humane Technology, quotes E.O. Wilson, who said that the real problem of humanity is that we have Palaeolithic emotions, mediaeval institutions and godlike technology. He talks about the *race to the bottom of the brain stem*, in which technology is designed to dominate our attention:

> 'Technology actually overwhelms human weaknesses. And I want to claim to you today that this point being crossed is at the root of bots, addiction, information overload, polarization, radicalization, outrage-ification, vanity-ification – the entire thing which is leading to human downgrading. Downgrading our attention spans, [our] relationships, stability, community, habits, downgrading humans.'[25]

In the interests of keeping us captivated by our screens for as long as possible, technology is designed to push our most primitive buttons, the ones that correspond to our basest instincts like gossip, envy, conspiracy theories and general craziness. Harris continues:

> 'With over a billion hours on YouTube watched daily, 70 per cent of those billion hours now are from the recommendation system. So the AIs are actually downgrading and overwhelming our free choice. So free will is being colonized.'

This is in service to profit, of course – not just through inviting us to click on adverts, but to obtain data about our mouse movements, click patterns, and eye gaze that can then be used to feed the algorithms that predict our future behaviour and offer to sell us things we didn't yet know we wanted. This data is enormously valuable in the right – or possibly wrong – hands.

Tristan Harris defines humane technology: 'an interface is humane if it is responsive to human needs and considerate of human frailties.' In other words, if it refrains from using our data to dominate and manipulate us, and instead genuinely makes our lives better.

Media

In 2021, I spoke with Charles Eisenstein, essayist, thinker, and author of *The More Beautiful World Our Hearts Know Is Possible*, for my podcast. I asked him how he's raising his 8-year-old son to be a man in the twenty-first century, and he shared some reflections on entertainment. Given that the main theme of our discussion had been about shifting from a Myth of Separation to a Story of Interbeing, he remarked that it's a challenge when movies like *Star Wars* portray goodies versus baddies, and violence as a solution.

> 'I try to provide some parental guidance, giving context. I'll say something like, "Wow, looks like they solve every problem with violence." [We] program people to see the world, to see progress, as a matter of overcoming – finding an enemy and overcoming that enemy. This is a lot of how geopolitics works, our political culture, our medical culture. We're very comfortable in a situation where there's an identifiable bad guy. It's a relief, because then you know what to do.'[26]

There is a natural human tendency to identify the in-group and the out-group, and by definition, whatever group you're in is the in-group, and most groups identify as 'good', making the designated out-group, by implication, 'bad'. As the saying goes, 'One man's terrorist is another man's freedom fighter'.[27]

It then becomes the mission of the 'good' in-group to conquer evil by prevailing over the 'bad' out-group. We are exposed to an infinite number of variations on this simplistic dynamic in entertainment media from childhood onwards, until it becomes a foundational assumption as to how the world works.

As Charles Eisenstein says:

> 'You attack something, you suppress something, you banish something, you kill something, you destroy something. That's our comfort zone, and that is culturally programmed. You win a war against radical extremists, against Islam, you wall off the immigrants, you find some identifiable source of all the problems, then you never have to look at yourself. And you have no access to the matrix of causes that includes yourself, that includes everything.'[28]

What about news media? The same as entertainment, but more so. Winners, losers, allegations, accusations, condemnations, confrontations, fights, wars, victories, defeats. Of course, this is a reflection of what is going on in the world, but it's a partial reflection, not a fair, full or balanced one. Although good and beautiful things are happening in the world, we wouldn't know it to look at the news. When news media depend for their survival on advertising revenue and the sale of user data, they have every incentive to blast out attention-grabbing headlines as quickly as possible in order to beat the competition, increase circulation, and hence get more income from advertisers. This tainted window on the world is generating excessive fear and loathing, and dividing populations against themselves.

Public debate can be a sign of a healthy democracy, but the danger is that debate gets trammelled by the media's framing into a narrow and heated debate that is largely irrelevant to the real issues. Noam Chomsky sounded a warning in 1998, and the advent of social media in the intervening years has only exacerbated the problem:

> 'The smart way to keep people passive and obedient is to strictly limit the spectrum of acceptable opinion, but allow very lively debate within that spectrum – even encourage the more critical and dissident views. That gives people the sense that there's free thinking going on, while all the time the presuppositions of the system are being reinforced by the limits put on the range of the debate.'[29]

Increasingly, we are seeing narratives being suppressed, cancelled or dismissed as disinformation, fake news or conspiracy theories. These allegations may sometimes be fair, but in many cases they are convenient labels for stories that run counter to the particular viewpoint or agenda of those with the power to control the media narrative. Journalists can't afford to annoy sponsors or shareholders, and sometimes find themselves on the wrong side of the fine line between press and propaganda.

Hannah Arendt, the Holocaust survivor and political philosopher, highlighted how deeply dangerous this is in a 1974 interview with the French writer Roger Errera:

> 'The moment we no longer have a free press, anything can happen. What makes it possible for a totalitarian or any other dictatorship to rule is

> that people are not informed; how can you have an opinion if you are not informed? If everybody always lies to you, the consequence is not that you believe the lies, but rather that nobody believes anything any longer ... And a people that no longer can believe anything cannot make up its mind. It is deprived not only of its capacity to act but also of its capacity to think and to judge. And with such a people you can then do what you please.'[30]

And so power over the press leads to power over the people.

Science and Medicine

The basic steps of the scientific method are:

a) make an observation that describes a problem;
b) create a hypothesis;
c) test the hypothesis;
d) draw conclusions and refine the hypothesis.

It all sounds very clean and clinical and honest.

But scientists are human beings, and are as prone to power plays as the rest of us. When your whole career has been based on one hypothesis and an upstart new theory comes along, defensive battle lines get drawn. Scientific debate should be a valuable technique for pursuing truth, but genuine debate can get crushed underfoot when one or both sides are determined, not to find truth, but to win. And so mud starts flying, science becomes dogma, and research gets suppressed, censored or defunded.

Medicine can be particularly contentious. A 2022 article in the highly respected, peer-reviewed *British Medical Journal* warned that evidence-based medicine has become an illusion 'corrupted by corporate interests, failed regulation, and commercialisation of academia'.[31] Rather than supporting health, the primary function of pharmaceutical companies is to focus on disease, diagnosis and drugs, a toxic trifecta when combined with the profit motive. Find a glimmering of a new 'disease', create medication to deal with it, fudge the clinical trials to exclude unfavourable data, issue a press release (to news outlets that you sponsor) to spread the bad news that there is a new affliction, and the good news that there is a miracle cure.[32] It is particularly alluring if the 'cure' comes in a convenient pill rather than requiring all the inconvenience of a lifestyle change.

Religion

Religion is, by definition, a matter of faith, but humans have a nasty habit of mistaking it for fact. Faith and reason make uneasy bedfellows, and debate over religious differences will rarely, if ever, reach a resolution. Meanwhile, theological differences have been used to justify crusades, witch hunts, missionaries, massacres and invasions since time immemorial.

The English philosopher Alan Watts wrote:

> 'Religions are divisive and quarrelsome. They are a form of one-upmanship because they depend upon separating the "saved" from the "damned," the true believers from the heretics ... All belief is fervent hope, and thus a coverup for doubt and uncertainty.'[33]

I could probably write about any of the organised religions here but, as the daughter of two Methodist preachers, I am going to choose Christianity because it is the religion I know best – with apologies to Christians for picking on them, and to all the others for not.

The great paradox of Christianity is that 'love thy neighbour as thyself' often gets overlooked in favour of 'no one comes to the Father except through me' (i.e. through Jesus). I am not criticising Christianity itself here, but rather those practitioners who hold the fundamentalist belief that theirs is the one and only true faith. This monotheistic attitude of 'our way is the only way' is problematic.

In *21 Lessons for the 21st Century*, Yuval Noah Harari suggests that polytheistic religions, by contrast, are literally a broad church.[34] While monotheism allows for only one true god, deeming it heresy to worship any other, polytheists are far more tolerant, accepting that different people will worship different gods. Polytheism allows religious freedom, while monotheistic religions such as Christianity and Islam have led to crusades, inquisitions, discrimination, persecution and execution.

I don't wish to criticise religion, which could and should be a thing of peace and harmony, but rather to point out that when religion is practised in a dominator paradigm, it becomes just another tool that would-be dominators can use to oppress the dominated in a hierarchy of control. The doctrine of original sin puts us in the wrong from the moment we are born. Promises of heaven and threats of hell cast us on the mercy of a capricious and judgemental god. Contact with the divine is mediated through a hierarchy of preachers, priests and popes. Prosperity theology, the belief that material success is a sign of divine favour, compounds the misery of those who have not fared well financially by telling them that not only have they failed in a meritocratic society, but their poverty is a sign of God's disapproval.

Small wonder that we seek comfort in materialism, hedonism, distraction and addiction in order to numb this existential loneliness, cut off from sources of spiritual consolation.

Complicity by the Dominated

Just before I wrap up this chapter, in which I have dipped a tentative toe into some massive topics, I need to say something about the dominated as well as the dominators. I can't speak as a person of colour, or from an LGBTQ+ perspective, or as a member of the many other groups that lose out in a dominator paradigm, but I can speak as a woman. So here are my thoughts on the specific form of domination that we know as patriarchy.

This is going to be controversial, but I say this to demonstrate the power of stories. The stories we tell ourselves about who we are might put the top dogs on top, but it can also put the underdogs under. If we spend our time getting angry at the privileged and powerful, we separate ourselves further from them. We take ourselves further away from claiming our own privilege and power.

To look at this in practical terms first, let's think back to the rigged game of Monopoly. Members of the dominating class believe they deserve to be there. They think they've earned it. It doesn't matter that it's obvious to everyone else that they just lucked out. From their perspective, they are entitled to their privilege.

They also identify as good people (mostly). So, if they are faced with an angry mob telling them that they don't deserve their riches and that they're evil, they're not going to respond well. They will only grip more tightly to what they have, and use their power to clamp down on those who criticise them.

What if, rather than griping about the dominators, we put our own house in order? If the story we're telling ourselves is, 'I want what they have', then we're emphasising our own lack of the very thing we want. The American novelist and activist Alice Walker wrote, 'The most common way people give up their power is by thinking they don't have any.'[35] Waiting to be invited to the table, or standing around whining about being powerless, doesn't achieve anything.

We know we all need to work together to create a better future, and we women have some serious work to do. This isn't about 'leaning in' (à la Sheryl Sandberg) to the masculine world, this is about women's inner journey, coming up with a better story about what it means to be born female at this time in history. Our culture may be biased against us, but we don't need to be against ourselves. Having been overlooked, ignored, insulted, derided or threatened for a long time now, it can be really hard to speak up, but we have to find a way.

The challenge here is that, having won admittance to the male bastions of the workplace and government, we've found ourselves in systems that primarily recognise and reward masculine ways of wielding power. Shifting paradigms is hard. Feminist and classics scholar Mary Beard writes:

> 'We have no template for what a powerful woman looks like, except that she looks rather like a man.'[36]

A woman will rarely, if ever, out-man a man, although clearly some have tried. Many women who try to succeed according to masculine metrics of success find it stressful and detrimental to their health and wellbeing as they try to tread the impossibly fine line between being effective and assertive, while not being perceived as handbaggers[37] or ball-breakers.

You may have heard of the psychological study at Columbia Business School, in which subjects were presented with a description of a successful executive, a real-life venture capitalist called Heidi Roizen. One group received the original version, featuring Heidi. The other group received an amended version, in which Heidi had been swapped out for a fictional male counterpart called Howard. Apart from the names and pronouns, everything else in the accounts was identical. The subjects rated both Heidi and Howard as equally effective, but while they found Howard to be a smart and likeable leader, they thought Heidi was too aggressive. Depressingly, even the women disliked Heidi.[38]

I'll come clean and admit that I've noticed this in my own perceptions. I don't much care for women who come across as cold, brusque and professional, while it probably wouldn't greatly affect my opinion of a man. I expect women to be kind as well as effective. If a man behaves like a git, I'm not surprised. If a woman does, I'm disappointed. I definitely have higher expectations of women than I do of men. And so I perpetuate the double standard that makes life so much harder for women.

Mary Beard throws a lifeline:

> 'We have to be more reflective about what power is, what it is for, and how it is measured. To put it another way, if women are not perceived to be fully within the structures of power, surely it is power that we need to redefine rather than women?'[39]

She is absolutely right – but who among us will redefine power? Men have the power, and there are some men who still see women's ascendancy as a zero-sum game – more for us is less for them – rather than noticing that patriarchy serves nobody. There are, though, many enlightened male allies. As the economist Tim Jackson said about the patriarchy when I interviewed him for my podcast:

> 'Men have suffered under this. There's a deep unhappiness beneath our vision of masculinity, and there's also a sense of freedom in moving away from it.'[40]

As Hillary Clinton (herself a casualty of resistance to powerful women) proclaimed, equality is about a lot more than women's rights. It's about human rights. And if we can't get human rights sorted out, there isn't too much hope for nature's rights. The way we do something could be said to be the way we do everything, and if we interact with the world from a mindset of domination, then we will keep looking for an 'other' to dominate.

No matter what the context, the dominator model boils down to the need of some people to believe themselves *better than*, which often springs from the insecurity of inwardly believing themselves *less than*, which in turn springs from *separate from*. If we saw ourselves as deeply connected, rather than fundamentally separated, what a different world this would be.

We need to reconnect with ourselves, to recognise our intrinsic wholeness. I notice how much contemporary self-help literature revolves around dividing our own self against itself – it's about identifying something wrong with ourselves and then overcoming, conquering or fixing it. We're told that we have a flawed self, and we're encouraged to find the better self that is going to beat the flawed self into submission. But who or what is it that we're defeating? Ourselves? Indigenous spiritual activist Juliet Diaz points out the fallacy of seeking to become what we already are:

> 'We are already enlightened beings, yet the world tells us to seek – seek enlightenment, seek beauty, seek wholeness, seek sacredness – when we are already all those things. The misunderstanding lies in the belief that we must attain something, when in truth, we must detach ourselves from the things that tell us we aren't whole.'[41]

As within, so without. If we're trying to dominate even ourselves, rather than treat ourselves with kindness and compassion, what hope is there that we will show kindness and compassion towards others?

In Part Four, I'll return to these themes as they might look in a different world – a world based on connection rather than separation, compassion rather than domination – but before we get there I'll share some thoughts on the deeper currents underlying the choppy waters of twenty-first-century civilisation.

A word of warning and reassurance: Part Two is going to get somewhat stormy, before we reach calmer waters in Part Three.

PART TWO

THE EVEN WORSE NEWS

'The intuitive mind is a sacred gift and the rational mind is a faithful servant. We have created a society that honours the servant and has forgotten the gift.'

Albert Einstein

Chapter 5

Psychology

'We cannot reason ourselves out of our basic irrationality. All we can do is to learn the art of being irrational in a reasonable way.'[1]

Aldous Huxley

I arrived at Yale in August 2012 as a frustrated environmentalist. The danger from climate change and our other environmental challenges seemed to me so clear and present that I couldn't understand why we weren't all running around as if our hair were on fire, trying to sort it out.

I was one of the sixteen Yale World Fellows in the 2012 cohort, chosen from a pool of around 2,000 applicants for Yale's prestigious international leadership programme. The one-semester fellowship exposed our multi-ethnic, multidisciplinary crew to a range of subjects, theories and perspectives. A parade of academic luminaries made their way to the Betts House mansion on Prospect Hill to enlighten us – from a professor of divinity to a four-star general, a presidential candidate to a Nobel Prize-winning economist.

Alongside the core curriculum of the World Fellowship, Fellows were encouraged to choose two or three additional classes to audit. I chose environmental writing, at Yale's renowned School of Forestry and Environmental Studies (as it was known then, now called the School of the Environment), and an undergraduate psychology class, simply titled 'Thinking'. The course description was equally innocuous: 'A survey of research findings and theories of how we "think", and their real-life applications.'

Note the quotes sceptically placed around the word 'think'. The course would indeed challenge my faith in the human capacity for rational thought.

Our classes took place in a small seminar room on the upper ground floor of the Department of Psychology at 2 Hillhouse Avenue, New Haven, a drab brownstone with a flight of wraparound steps leading to a front door scuffed by the passing traffic of generations of students. Our professor was a Korean woman called Woo-kyoung Ahn, whose illuminating presentations were interspersed with often amusing video clips to illustrate key points. Her Thinking class now attracts around 500 students per semester, so I'm lucky I beat the rush. There were around twelve of us in the class when I took it.

Week after week, I sat in that dingy little room with undergraduates less than half my age, feeling as if the secrets of the universe – or at least, the secrets of the human mind – were being revealed to me, and wondering how different my life might have been if I'd known this sooner. The main purpose of the course was to highlight the quirks of our cognitive processes. We studied a broad range of logical fallacies and thinking errors such as confirmation bias, causality versus correlation, availability heuristics, base rate neglect, anchoring, priming, attentional bias, neglect of probability and hindsight bias.

I had fondly believed that humans were rational beings. The Thinking class, if not exactly blowing that belief out of the water, revealed that our rationality is extremely limited. The world started to make a lot more sense – or at least, I started to understand why it so often *doesn't* make sense.

Our Lazy Brain

Human irrationality largely arises from the fact that the brain is power-hungry. By that I don't mean that it's a megalomaniac, although that's definitely true in some cases. I mean, it uses a lot of energy.

The average brain accounts for about 2 per cent of our bodyweight, but it consumes about 20 per cent of the body's calorie budget. The brain's preference for shortcuts is often explained on the basis that historically every calorie used was a calorie we had to find by hunting or foraging, so evolutionary biology selected for brains that delivered more bang for the calorie-buck. The more preconceptions, or heuristics, we have about reality, the less work the brain has to do, the fewer calories it requires, and the more likely our ancestors were to survive times of hunger. Nature rewarded efficiency over accuracy.

As it turns out, you're unlikely to lose much weight by thinking extra-hard because the difference is marginal: reading *A Brief History of Time* will only burn about 5 per cent more calories per hour than reading *Hello* magazine. But anyhow, that's our story about why our brains take the path of least resistance, and we're sticking to it.

The first thing we learned during Thinking was how narrow our beam of attention is. Our professor showed a video of people throwing a basketball, asking us to count how many times the players in white vests pass the ball. You may well have seen this experiment – it's been around for a while. We dutifully watched the video, counting the passes. Afterwards, Professor Ahn asked if we'd noticed anything unusual.

[Spoiler alert: skip the next paragraph if you want to watch the video and do the experiment for yourself.[2]]

Turns out, even bright Yale undergrads do no better than the average: over half of all participants fail to notice a man dressed in a gorilla suit walk into the frame, pause in the middle of the screen to thump his chest, and then exit stage right. (I sexistly assume it's a man – hard to tell, because of the gorilla suit.) They are too busy paying attention to the players in white and the passing of the ball.[3] The non-gorilla-noticing participants are astonished when they are told what they missed, and will often say they couldn't possibly have failed to notice such an obvious anomaly. And yet ... they did.

The key point here, which will come up again later, is that our attention is a rather poor and feeble thing. Our conscious mind can process fifty bits of information per second, compared with the stunning 11 million bits that our subconscious can handle,[4] so we have to be very selective where we place our narrow beam of conscious attention.

According to developmental psychologist Alison Gopnik, consciousness narrows as a function of age. Infants have what she calls a *lantern of attention* – a diffuse, peripheral awareness – while adults with a fully developed prefrontal cortex have a *spotlight of attention*[5] – a tightly focused beam that shines on the specific object at hand. Attention is a zero-sum game; the more attention we put on one specific thing (white vests and counting ball passes), the less attention there is for anything else (random guys in gorilla suits). This *selective visual attention* is the secret of most close-up magic tricks; the magician masterfully directs your spotlight of attention over here, while over there he or she deftly does whatever they do to make the rabbit disappear. Incidentally, politicians employ a similar sleight of hand to bury unfavourable news while the public's attention is distracted elsewhere.

Our brains trick us into thinking that we see and know far more than we actually do. They also kid us that we are more rational than we are. And even worse, when our irrationalities are pointed out to us, we tend not to believe it. We have a blind spot about our blind spots. As Adam Rutherford writes in *The Observer*:

> 'It's called "bias blind spot": the inability to spot your own biases, coupled with a readiness to identify them perfectly well in others. I definitely don't have this one at all. But you almost certainly do.'[6]

In short, our cognitive biases, limited attentional capacity and blind spots mean we mostly believe our stories about what is happening, and our stories are demonstrably unreliable.

Cognitive Biases and Climate Change

Many of the themes from the Thinking course appeared in George Marshall's book, *Don't Even Think About It: Why our Brains are Wired to Ignore Climate Change*.

George lives in Wales, but was on a flying visit to Yale and I was able to finagle a few minutes with him by offering to escort him across campus from his meeting with Anthony Leiserowitz at the School of Forestry and Environmental Studies, to a meeting with Dan Kahan at the Law School. I can be quite shameless when there's someone I'm keen to talk with and they're on a tight schedule.

George is the founder of Climate Outreach, and his specialist subject is how to talk about climate change with deniers and the disengaged in ways that are respectful and effective. I arrived at Tony Leiserowitz's office to find a mild-mannered, bespectacled man with his coat already on, his trademark fedora hat set at a rakish angle. I made the most of our short time together by peppering him with questions, which he bore patiently.

It seemed appropriate that I caught George in transit from the director of the Yale Program on Climate Change Communication (YPCCC), which produces the Six Americas report[7] on the six distinct responses of the American public to climate change, to the recognised expert on cultural cognition, which examines the way that individuals assess risk not according to the actual seriousness of the risk, but, rather, according to their cultural conditioning. George's book sits at the intersection of Tony's and Dan's work, exploring in detail these themes of cognitive biases, enculturation and communication, setting out all the ways in which humans are poorly designed to perceive long-term and largely invisible threats. I recall a sense of growing despondency as I read it. Each chapter seemed to reveal more reasons why we are never going to wrap our quirky minds around climate change in a way that will lead to action.

George concludes with a relatively upbeat chapter on how climate campaigners can take these cognitive biases into account and find workarounds to galvanise change. But I'd given up hope by then. Given all the psychological obstacles described in the earlier chapters, I'd already reached the conclusion that I'd been hopelessly naïve in my assumption that all I needed to do was to say, 'Hey, folks, we have a problem – and it's a very big and existential one, so maybe we should do something about it?' It was clearly a lot more complicated than that, and our biggest challenge lay not in the Amazon, Arctic, or Pacific Ocean, but inside our own heads.

Confirmation Bias

We edit, distort and delete information in order to confirm what we've already decided, and ignore any contrary information. There might be people, for example, to whom you took an instant dislike, and from then on they could do no right in your eyes. Everything they did was viewed through your lens of distaste.

Confirmation bias explains how two groups of people holding opposing views can look at the same set of facts and find confirmation for their point of view. Think of creationists versus evolutionary biologists, flat-earthers versus round-earthers, Brexiteers versus Remainers, Democrats versus Republicans, and any conspiracy theorist versus the rest of the world. Same reality, different perspectives.

Often, this leads to judgement. I'll use myself as an example: when I found out about climate change, it seemed obvious to me that this was the most important issue facing humanity and something must be done immediately. I set about changing my lifestyle. Then, looking around at other people, I concluded from their behaviour they were clearly oblivious to the problem. Once they were aware, surely, their behaviour would change, as mine had done. But to my surprise and frustration, it completely didn't. I fell into the classic *uninformed, stupid, evil* trap, which goes like this:

> If people are not seeing reality the way I see it, it must be because they are ill-informed. I shall therefore inform them, and then they will agree with me.
>
> If I inform people and still they don't agree with me, it must be because they are stupid. But what if I find evidence that they are actually smart, yet are wilfully ignoring what seems to me a very real danger?
>
> I then conclude that they must be evil; if they continue to act recklessly towards our ecosystems and they are neither uninformed nor stupid, they must be maliciously persisting in making choices that are incompatible with the wellbeing of future generations.

But they're *not* evil, nor stupid, nor ill-informed. They are just shining their spotlight of attention on different facts, or looking at the same facts and arriving at a different conclusion based on their prior conditioning.

Even scientists, supposedly the most rational of all humans, are susceptible to confirmation bias. Let's look again at the scientific method, which starts with formulating a hypothesis. The problem is that the hypothesis will determine the design of the test and what instruments are selected to measure the result. Too often, scientists set out to *confirm* their hypothesis, when a genuinely open approach would include trying to *disconfirm* it. The instruments they use to measure their results will be appropriate to the results they expect: if they're looking for light, they will measure for light, and might not notice magnetism or sound or heat. As the philosopher-psychiatrist Iain McGilchrist says, 'The model we choose to use to understand something determines what we find.'[8]

Similarly, peer-reviewed research isn't all it's cracked up to be. Fellow scientists tend to give positive reviews of research that tallies with what they themselves believe, rather than intrinsically good science. Larry Dossey describes this bias in *One Mind*:

> 'What gets denied publication in professional journals sometimes has little or nothing to do with the veracity of the data itself. Decisions about the awarding of research grants and the publication of papers often appear to be made by what's been called the GOBSAT method – good old boys sat around a table.'[9]

He quotes James Watson, who won a Nobel Prize for his work on the structure of DNA. Watson declared, 'One could not be a successful scientist without realizing that, in contrast to the popular conception supported by newspapers and mothers of scientists, a goodly number of scientists are not only narrow-minded and dull, but also just stupid.'[10] (Presumably he counted himself among the 'successful' scientists rather than the stupid ones – so maybe he had his own confirmation biases.)

Scientists are as human as the rest of us and share our general unwillingness to change our minds once they're made up. If forced to consider contradictory evidence, we tend to double down and believe all the more zealously – witness political partisanship.

The sad irony is that even once we're aware of confirmation bias, we seem to be helpless to overcome it. We're naturally more comfortable with the worldview we know, and find it hard to open our minds to something that may be closer to the truth.

Causality Versus Correlation

I brush my teeth every morning, and the sun rises every morning, but I would be sorely mistaken if I think my dental routine causes the sunrise. Just because two events take place close in time, and usually in the same sequence, doesn't prove that the one causes the other.

A much-cited example is the correlation between annual deaths by drowning in swimming pools and the number of Nicolas Cage movies that came out that year.[11] It would obviously be ludicrous to hold Mr Cage personally responsible for these untimely deaths, and yet when we think we see a pattern emerging we often leap to incorrect conclusions in ways that are only marginally less ludicrous. Biology has selected for the useful predictive abilities of brains that are good at spotting patterns; if the last three folks to eat that kind of mushroom got sick and died, it's probably best to find something else for lunch. We love to think we've figured out the connection so we can use it to our advantage. Know anybody who helps their favourite team win by wearing special lucky underpants?

But we often get it wrong. There might be no connection at all between the events, or we might be mistaken about which is the cause and which is the effect, or there could be a third factor that causes both the observable happenings. This has been the crux of the argument put forward by many climate change deniers: we have increasing carbon dioxide and we have increasing temperatures, but they dispute the causal link. They tend to focus instead on the correlation between climate-related research and successful funding applications – in other words, cynically suggesting that climate scientists subscribe to the view that anthropogenic carbon dioxide emissions cause warming so they can get their research funded.

The more complex the system, the harder it becomes to identify clear chains of causality. When it comes to global climate, the timespans are much longer and the data sets more complicated than football victories and lucky underwear.

Availability Heuristics

Over the last fifteen years, an average of just under 500 people per year die in plane crashes globally. This is tragic for those people and their families, but is not a good reason to be afraid of flying. Around 1.3 million people die annually in car crashes, 400,000 of malaria and 12,000 die after falling down the stairs. And 659,000 die of heart disease ... in the US alone.

The term 'availability heuristic' was coined by the Nobel Prize-winning duo Daniel Kahneman and Amos Tversky, for the unconscious assumption that 'if you can think of it, it must be important'. Things that come to mind more easily are believed to be more common than they actually are. The availability heuristic can form an unholy alliance with the media's over-reporting of sensational and disturbing news to make us believe that child abductions, murders and shark attacks are happening all over the place, when thankfully they are relatively rare.

On the flip side, widespread reporting of huge lottery wins can make us believe we might get lucky, when in reality the chances are vanishingly small. Just so you know, the odds of winning the US Mega Millions lottery are 1 in 325 million. If you're American, back-of-the-envelope pundits reckon your chances of becoming president are 1 in 10 million,[12] so recalling the potential earning power of ex-presidents, and the more favourable odds, statistically you'd be better off launching a presidential campaign than buying a lottery ticket – although bear in mind there's also a 1 in 11 chance you'll be assassinated while in office.

The point is that we tend to overestimate the frequency and severity of risks that easily come to mind, while underestimating risks that might be more frequent or more severe, but are not sufficiently sensational to hit the news headlines. We're more likely to pay attention to stories that seem urgent, well defined and personally relevant. Most ecological disaster stories, unfortunately, emerge slowly, are complex, and if we live in the Global

North they mostly affect people we don't know and who look different from us. This is a problem.

A corollary of availability heuristics is the peril of shifting baselines – or shifting waistlines, as Daniel Pauly joked during his talk at TED Mission Blue, looking ruefully at a slide of his younger, slimmer self. Like the pounds that stealthily accumulate around our middle over the years, we tend not to pay attention to gradual changes like the slow but dramatic drop in the number of songbirds, or hedgehogs, or insects splattered on the car windscreen. It's only when we find we have to go up a jeans size, or read a news headline that flying insects have declined by over 75 per cent in the last twenty-seven years,[13] that we realise how change has crept up on us.

In the modern Western world at least, we tend not to be good at paying attention to long timespans. Nobody ever made a blockbuster about a slow-burning crisis that the hero headed off with plenty of time to spare. We've been conditioned to prefer our crises urgent and exciting, with the hero saving the day a split second before the countdown clock reaches zero. The biggest crises of our times lack Hollywood-style drama, which is an unfortunate marketing issue.

Hindsight Bias

Once we know something, we tend to act as if we have always known it, and regard anybody who doesn't yet know it as an ignoramus. This can also lead us to judge others – or ourselves – harshly, forgetting that we didn't know then what we know now, so a decision had to be made using imperfect information.

Maybe you have one of those annoying friends who enjoys saying, 'I knew that would happen' – although it's funny how rarely they share this valuable insight when it might actually have been useful. Or maybe on the news you've seen the neighbour of a serial killer saying they always thought he (or very occasionally, she) was rather odd – but apparently never thought to mention this to the police.

Research identifies three main factors in hindsight bias:

People tend to forget their incorrect predictions, and remember the correct ones (ref: lucky underpants).

When the event, in retrospect, seems predictable, we like to think we predicted it.

People feel more secure if they believe the world is a predictable place.

The danger of hindsight bias is that it might lead us to think we're better at predicting the future than we actually are. Research suggests we're not very good at all, even (or maybe especially) when it comes to predicting our own future feelings, as psychologist Dan Gilbert points out in his TED Talk:

> 'At every stage of our lives we make decisions that will profoundly influence the lives of the people we're going to become, and then when we become those people, we're not always thrilled with the decisions we made. So young people pay good money to get tattoos removed that teenagers paid good money to get. Middle-aged people rush to divorce people who young adults rushed to marry.'[14]

If we're not good at foreseeing consequences but we think we are, that's a dangerous combination. We tend to barge ahead in an over-confident fashion, thinking we know what we're doing. The evidence – think of fossil fuel extraction, habitat destruction or introduction of non-native species – indicates that many of the ideas that seemed so good at the time turn out not to be. While a few prudent souls call for the precautionary principle – *maybe we should do a small pilot scheme, let it run a while, see how it goes* – humanity's prevailing attitude seems to be: *woohoo! what can possibly go wrong?*

Bystander Effect

You've probably heard horrifying stories of victims being hit by cars, suffering seizures or even being murdered in broad daylight, while nearby people do nothing to help. Even if people see what is happening, we have a tendency to pick up on the responses of others. If nobody else is taking action, we tend to assume there is nothing to worry about.

The bystander effect is also known as bystander apathy, and in a world where so many seemingly pressing issues vie for our attention, we could be forgiven for suffering apathy arising from compassion fatigue. We can become numb to problems that don't affect us personally.

It's summed up in the words of German Lutheran pastor Martin Niemöller about the cowardice of some German intellectuals and clergy during the Holocaust.

> 'First they came for the socialists, and I did not speak out –
> Because I was not a socialist.

Then they came for the trade unionists, and I did not speak out –
Because I was not a trade unionist.
Then they came for the Jews, and I did not speak out –
Because I was not a Jew.
Then they came for me – and there was no one left to speak for me.'[15]

Because we're susceptible to taking our cues from those around us, social influence has enormous power over our habits as consumers. It can be hard to inconvenience ourselves or make sacrifices when we see friends and neighbours carrying on much the same as always. Why should we stop flying, or resist using a global online shopping giant, or avoid imported food when everybody else is doing it? It takes a particularly strong moral backbone to resist social norms in favour of a more ethical lifestyle.

Authority Bias

Closely related to the bystander effect is authority bias. In both cases, we abdicate our sense of responsibility to somebody else – bystanders, or people who we regard as having some form of authority.

Stanley Milgram was a psychologist at Yale in the 1960s. He's famous for the experiments he conducted in search of clues as to how the Nazi atrocities could have happened. A member of the public would respond to a newspaper advert asking for male volunteers for a learning experiment. On arrival at Milgram's lab, lots would be drawn to find out which man would be the 'learner' and which would be the 'teacher', but the draw was fixed so that the volunteer would always be the teacher, while Milgram's colleague would be the learner.

The learner would be taken into a room next door to have electrodes attached to his arms, while the volunteer would be invited to sit at a control panel that would administer electric shocks, with a row of switches ranging from 15 volts (labelled Slight Shock) to 375 volts (Danger: Severe Shock) to 450 volts (ominously labelled XXX). The learner would be given a list of word pairs to learn, after which the volunteer would test him by giving one of the words, and the learner was supposed to respond with the other half of the pair. If he answered incorrectly, the volunteer was told to deliver an electric shock, increasing the level of the shock each time the learner got it wrong. The volunteer would be able to hear the (pretend) wails of anguish from the room next door as he administered ever-stronger shocks.

If he hesitated, the experiment supervisor – the figure of authority – would urge him to continue.

Two-thirds of the volunteers went all the way to the highest level of electric shock. All the participants went as far as Severe Shock. Eighteen variations on the experiment yielded similarly shocking (so to speak) results. As Milgram noted in a 1973 article in *Harper's Magazine*, when the subjects' natural human aversion to hurting others ran slap bang into blatant authority, authority usually won.[16]

Milgram's analysis explains that people generally have two states of behaviour: the autonomous state, in which they take responsibility for their actions, and the agentic state, in which they allow others to direct them, believing that the responsibility lies with those doing the directing: *I was only following orders*. He argued that ordinary people, conditioned to follow orders, can become weaponised as agents of terrible destruction. Even when they can hear the screams, and in some sense know that they are acting contrary to their deepest moral standards, few people have the inner resources to defy harmful instructions issued by an authority figure. Authority usually trumps morality.[17]

People are more likely to enter the agentic state when they perceive the authority as being legitimate, and perceptions of legitimacy can be influenced by the authority figure wearing a uniform or (as in the Milgram experiments) a lab coat. This strategy is used in advertising and many other contexts today – would you be more inclined to buy vitamin supplements, medications or skin creams from people in lab coats or people in ordinary clothes?[18]

Sunk Cost Fallacy

Ever waited and waited for a bus, telling yourself, *I've waited this long – it must be just around the corner*? Or gone to the cinema and watched the film to the end, even though you realised in the first five minutes you didn't like it? Or stayed in a job or a relationship long after you knew it wasn't working for you?

It would be a rare mortal who hasn't, at some time or another, fallen into the sunk cost fallacy.

Part of the problem is that Western culture tends to value tenacity and determination, and it can be hard to distinguish between three very different things: fortitude, faintheartedness and fallacy. It's one of the life skills

I still find tremendously challenging, even after my half-century of lived research: when is it right to quit? And when might success be just around the corner, like that elusive bus? When am I being commendably resolute, and when am I being pointlessly stubborn?

We're no better at it collectively than we are individually. The banks that were *too big to fail* during the 2008 global economic crisis; the old and obsolescing industries that are propped up by governments that don't want their unemployment figures to look bad; the entire capitalist enterprise built on foundations of cheap labour, environmental exploitation, pollution and consumer manipulation.

Sunk cost is closely related to another fallacy – appeal to tradition, or *we've always done it this way*. But things change, and what worked in the past may not be fit for purpose any more. It's increasingly obvious that many of our systems are failing, but so much has been invested in them that it would take a bold leader indeed to suggest we rip them up and start again with something better suited to current realities.

Status Quo Bias

Following on from the sunk cost fallacy, humans tend to prefer the devil we know to the devil we don't – or even the angel we don't. Even when a change would clearly be for the better, we tend to stick with the familiar rather than venture into the unknown.

Again, we can see why evolutionary biology might favour less adventurous souls. Doing something that has never been done before comes with risks. Behavioural economists Daniel Kahneman and Amos Tversky found that people feel greater regret for bad outcomes resulting from new actions than for bad consequences resulting from inaction – so we often default to doing nothing. This is compounded by our general Western state of cognitive overload. As we are increasingly overloaded with options, sticking to what has worked in the past is a less difficult decision.

We tend to assume that defaulting to doing what we've always done is the safest bet. It hasn't killed us in the past, the reasoning goes, so it's unlikely to kill us now. But this assumption ignores the cumulative effect of doing what we've always done. Too much of anything, even a good thing, can become a bad thing. Persistently sticking with the known can take us into the unknown. Continuing to do what we think is safe can make us extremely unsafe – can, indeed, threaten life on Earth.

The G.I. Joe Fallacy

A-ha! you might think. But now we know about these cognitive pitfalls, we can take countermeasures to avoid them!

Sadly, it would appear that this is not so. Laurie Santos and Tamar Gendler coined the term 'G.I. Joe fallacy' to refer to the misguided notion that knowing about a thinking error is enough to overcome it:

> 'The name of this fallacy derives from the 1980s television series *G.I. Joe*, which ended each cartoon episode with a public service announcement and closing tagline, "Now you know. And knowing is half the battle." Santos and Gendler (2014) argued that for many cognitive and social biases, knowing is much less than half of the battle. They further argued that many people believe this is not the case, that knowledge is in fact enough to behave better, hence the G.I. Joe fallacy.'[19]

So not only do we fall into thinking errors with depressing frequency, but the belief that knowing about these errors will help us safeguard against them is in itself a thinking error. Are we stuck in an infinite regression loop?

We may think we see reality as it is, but these cognitive fallacies are just a few examples of the countless ways in which our brains get it wrong. I'll recap.

Confirmation bias leads us to filter information to confirm what we think we know, and disconfirm what we don't want to know. *Mistaking correlation for causation* makes us think we know why something happened, and can therefore predict when it will happen again in the future, but it's possible we've found a causal pattern where there is none ... and confirmation bias makes us ignore the times when our imaginary causal relationship failed. The *availability heuristic* also tends to make us more aware of the positives (when our pet theory worked) than the negatives (when it didn't), because the brain is much better at noticing presences than absences, hits rather than misses.

Even though our predictive capacities are demonstrably poor, *hindsight bias* gives us the illusion that we predicted something when we didn't, and are therefore likely to make good future predictions. The *bystander effect* means we take our cues from other people, and if they aren't panicking, then we don't, even when we should be. The *sunk cost fallacy* and *status quo bias* gang up together to keep us glued to an ongoing course of action, failing to notice when we become victims of our own success. And the *G.I. Joe fallacy* makes us think we can overcome these biases once we're aware of them – yet we don't.

In short, we're nowhere near as smart as we think we are. We're not even smart enough to know how not-smart we are, and so we blunder on with cocksure confidence when caution and humility might be more appropriate.

Is there really no escape? By the end of my semester at Yale, I had concluded there was not, and nor was there much hope we would wake up to the urgent necessity for environmental action. We might think we're changing things, but we're really just tinkering around the edges. We might think we're brave, but we're actually very reluctant to let go of what we know.

Change is hard, even if people want it, and they mostly don't. Part of the reason is external. Practically all of us are embedded in structures, especially economic and social structures, that enshrine the behaviours and attitudes that have worked well in the past. These cultural patterns run deep. As the business management expert Peter Drucker allegedly said, 'Culture eats strategy for breakfast.'[20]

And part of the reason is internal. Psychological research provides us with ample evidence that change is hard for the majority of people[21] (not that we needed research – we only have to look at the annual charade of New Year's resolutions to know this), even when there are very obvious and immediate benefits to them personally – and unfortunately most of the changes that the environmental movement has asked people to make have, if anything, appeared more like sacrifices. It is maybe one of the biggest failures of environmental messaging that it has focused on *save the whale* or *save the planet*, when an appeal to *save the humans* might have been more effective.

It doesn't help that we're all complicit. Many environmental communications arouse feelings of fear, guilt, shame or despondency, leading to responses ranging from denial (it's not happening), low prioritisation (it's less important than the economy), conspiracy theories (it's a hoax), finger-pointing (that country emitted sooner/more), cost/benefit analysis (it will cost our company/country too much) and resignation (it's too late anyway). It has been an ongoing challenge to generate enough concern to galvanise action, while not generating so much that we become frozen with fear.

The real kicker in all of this is that ecological degradation, like rust, never sleeps. All the time that we are bickering, dithering, denying and procrastinating, habitat keeps disappearing, ice caps keep melting and species keep going extinct. Nature isn't waiting for us to get our act together.

Chapter 6

Hemispheres

> 'The only certainty, it seems to me, is that those who believe they are certainly right are certainly wrong.'[1]
>
> Iain McGilchrist

In 2016 I was onstage at a conference in Cumberland Lodge, a Gothic house of impressive proportions picturesquely nestled into the rolling green landscape of Windsor Great Park. I'd originally expected to be in the audience – a friend had recommended the work of Iain McGilchrist, and I'd discovered to my delight that he was due to speak just a few miles from where I was then living in central Windsor, so had booked tickets.

But my email signature, describing myself as 'Author, Speaker, Ocean Rower', had come to the attention of the principal of Cumberland Lodge, Edmund Newell, who was writing a book on spirituality and the sea.[2] He wanted to hear more about my ocean rowing experiences and, after an initial conversation over coffee, he invited me to participate in a panel at the event.

The conference was called The Stifling Hand of Control, headlined by Margaret Heffernan, author of *Wilful Blindness*,[3] and Iain. I was already familiar with Iain's central thesis, presented in his seminal work, *The Master and His Emissary*.[4] He's concerned that the left hemisphere of our brains, with its reductionist worldview, is coming to dominate our society at the expense of the more subtle and intuitive right hemisphere. Iain fears that the shift towards the left hemisphere is becoming a self-reinforcing feedback loop, or *hall of mirrors* as he calls it, due to the left hemisphere's excessive self-confidence and its apparent obliviousness to the crucial role of the right hemisphere.

Other speakers included high-ranking individuals from the military, education, clergy and police. All presented different aspects of the same story: that our culture is becoming increasingly dominated by the left-hemisphere, with its love for control manifesting as a bewildering multiplicity of tiny rules dominating everyday life – hence the name of the conference.

I am honoured to now count Iain as a friend, and have stayed with him several times at his home on the Isle of Skye. He is an excellent host with a fine wine cellar, and we have enjoyed long evenings of conversation by an open fire. Whisky may, on occasions, be involved.

Iain disdains crude dualistic characterisations of the two hemispheres, so I hope he will forgive my simplistic summary here. For the definitive version, I urge you to read *The Master and His Emissary*, or the abbreviated version in *The Divided Brain and the Search for Meaning.*[5]

The basic facts are that the human brain, and the brains of most other animals, are made up of two hemispheres. The left hemisphere is good at linear and reductionist thinking, categorisation, logic and analysis, mechanical concepts and tasks requiring focused attention. It is optimistic, individualistic, has the monopoly on verbal language (which is why people who have suffered a left hemisphere stroke lose their ability to speak, but those with a right hemisphere stroke don't) and is blessed with a disproportionately robust perception of its own abilities. Some people regard it as the more masculine of the two hemispheres. I can't imagine why.

The right hemisphere is in many ways the opposite, the yin to the left hemisphere's yang. It is good at conceptual and holistic thinking, imagination, intuition, compassion, prefers the organic to the mechanical and sees the wider context. It is the seat of most emotions, apart from anger, which is the prerogative of the left hemisphere. It has no verbal language, has a high tolerance for paradox without needing to plump for one option or the other, and tends to be pessimistic, melancholy and doubtful. Some people think of it as the more feminine hemisphere. Again, I'm saying nothing.

Put another way, the left hemisphere is good at grasping and manipulating specific objects, while the right understands the relationships between things, and between us and our world. The left hemisphere controls the right hand, which for right-handed people is usually the hand that reaches out and takes hold of things, moves things, or pulls things apart. The left hemisphere also governs language, and we can see the parallels: words are not the same as the thing they describe, but as symbols for those things they enable us to grasp and manipulate ideas in a way similar to how hands grasp and manipulate objects.

Our best theory at the moment is that the two hemispheres were a positive evolutionary adaptation that allowed creatures to maintain tight focus on the task at hand (like eating their lunch) while also maintaining awareness of their surroundings (thereby not becoming someone else's lunch).

Ideally, the two hemispheres operate in harmony, each playing to its strengths. But here we run into the snag that inspires the title of Iain's book. The story, originally referenced by Nietzsche, goes that there was once a wise and spiritual master who had a small and prosperous domain. As his empire grew in size, he trained a number of trusted emissaries to oversee the welfare of its far-flung outposts. In his wisdom, he chose not to micro-manage his emissaries, but to let them get on with the job. Unfortunately, his most ambitious emissary, far from home and without supervision, began to abuse his power in order to further his own wealth and influence. 'And so it came about that the master was usurped, the people were duped, the domain became a tyranny; and eventually it collapsed in ruins.'[6]

A Balanced Brain – or Not

Albert Einstein, according to a 2013 study, was blessed with an unusual corpus callosum (the connective pathway between the brain's hemispheres). It had more extensive connections between certain parts of the brain compared with the brains of control groups.[7] We can't know for sure that these information superhighways are what made Einstein a genius, but they certainly didn't do him any harm. In ordinary mortals the corpus callosum has been getting narrower over time, increasing the separation of the hemispheres.

Einstein hinted that he preferred to dwell in the metaphorical, image-based world of the right hemisphere than in the verbal world of the left, which may offer a clue as to how he was able to imagine possibilities beyond the ken of most:

> 'I very rarely think in words at all. A thought comes, and I may try to express in words afterwards.'[8]

In his *Autobiographical Notes*, Einstein relates a vision he had as a 16-year-old in which he was chasing a beam of light. This led him to the theory of special relativity. His vision was intuitive and visual, rather than logical and verbal, so would seem to have originated in the right hemisphere.

So, if we want to be more like Einstein, the right hemisphere should act as the chief strategist, given its wider and wiser perspective, while the left hemisphere plays to its strengths as administrator and executive, effectively carrying out the orders it receives from the right hemisphere.

In a perfect world, the two hemispheres form a balanced and complementary double act, like Fred Astaire and Ginger Rogers (remembering that Ginger did everything Fred did, but backwards and in high heels – and indeed, in some ways the right hemisphere, like Ginger, has the more challenging role, but gets less credit for it). It seems likely that this equitable balance used to be the case, or at least, more so than it is now. However, given its ebullient self-confidence, the left hemisphere has gradually usurped the power of the right. The right hemisphere knows it needs the left, but the left has forgotten the right even exists.

Once I started to see the world through this lens of left-hemisphere dominance, I saw evidence of its reductionist, materialist perspective everywhere: billionaires wanting to cryogenically preserve their heads when they die, pending the day when the contents of their neural networks can be uploaded into a computer;[9] prioritising quantifiable 'friends' on Facebook over quality time with real-life friends; 'teaching to the test' diminishing the autonomy of teachers; compulsory cheerfulness at work;[10] a world increasingly run by algorithms.[11] The news media also show left-hemisphere tendencies – sensationalised headlines, simplistic perspectives, celebrities and public figures reduced to cartoonish caricatures of themselves with precious little subtlety or context.

The heart of the problem, as I understand it, is that the left hemisphere loves knowledge, facts, science, technology and progress. These are all good things, and our society correctly prizes them.

However, we should also prize wisdom, which is the superpower of the right hemisphere. Wisdom is the combination of intuition, context and the humility to reflect on our experiences to evaluate what worked and what didn't. These reflections in turn inform future intuitions about how best to proceed, given not just the immediate question, but also the broader implications. As the saying goes, 'Knowledge is being aware of what you can do. Wisdom is knowing when not to do it.'

Knowledge wants to hurtle forwards regardless, in pursuit of ever more knowledge. Wisdom wants to exercise the precautionary principle (counterbalancing hindsight bias, which is a particularly left-hemisphere-style fallacy). The left hemisphere is bold, modernistic, ambitious. The right hemisphere respects time-honoured ways of living.

Neither is all good, neither is all bad. It is all about balance.

Reduction and Reductionism

The left hemisphere's reductionist thinking spawned the view that we have only five senses. Aristotle's *De Anima* might have been the start of the idea that each sense had to correspond to an observable physical organ, and that other ways of knowing should be relegated to the woo-woo land of sixth sense and extrasensory perception.[12] There have been powerful incentives over the centuries for those who have such gifts (arguably, all of us) to suppress them. During the European witch hunts between the mid-fifteenth and mid-eighteenth centuries, people (mostly women) who were skilled in ancient healing practices were liable to be tried and burned as witches. Even those who escaped persecution would be powerfully motivated to hide their talents and not pass them on to daughters or apprentices, leading to a tragic loss of wisdom traditions and inheritable sensory capacities.[13] Elsewhere in the world, indigenous witch doctors and shamans have been dismissed as primitive savages.

One of the core tenets of the Age of Enlightenment (maybe ironically named), was that the evidence of our senses should be our sole source of knowledge; empirical evidence took precedence and intuition had to resort to sneaking around in the shadows. Even if they dare to follow in Einstein's distinguished footsteps by using intuition, modern scientists would rarely admit such a thing for fear of losing credibility, preferring to perpetuate the myth that they arrive at discoveries through logic alone.

Is it possible that humans used to have a greater intuitive sense of the underlying nature of reality, but the materialist worldview of the left hemisphere is increasingly blocking our ability to perceive invisible realms that exist beyond our paltry five senses? Modern science might tell us that what we see is all there is, but as neurosurgeon Paul Kalanithi points out in his beautiful contemplation of life and death, *When Breath Becomes Air*, science should be a subset of metaphysics, rather than the other way around:

> 'To make science the arbiter of metaphysics is to banish not only God from the world but also love, hate, meaning – to consider a world that is self-evidently not the world we live in.'[14]

Given the tiny fraction of all incoming data that the conscious mind can process,[15] the more our left hemisphere enforces its (literally) narrow-minded worldview, the more we cut ourselves off from valuable sources of information. According to Iain, we have paid dearly for our increasingly left-hemisphere orientation:

> 'We know so much, we can make so much happen, and we certainly invest much in the attempt to control our destinies. And yet, if we are honest, we feel as though it ought somehow to have added up to – more than this.'

He draws the connection between left-brain dominance and our disrespect for nature, and also correlates it with the rise of materialism, which in turn leads to unhappiness and mental illness. As a former psychiatrist, he's well qualified to know.

> 'Around us we can scarcely fail to see the evident global degradation and destruction of what we now call "the environment", but which is nothing less than the living world; the breaking up of complex, close-knit communities, and their ways of living in harmony with nature, that took at least centuries, if not millennia, to form; the substitution of a way of life that we have already determined in the West to be lacking in meaning, often aesthetically barren, driven by commercialism and morally bankrupt, devoted to the pursuit of pleasure and happiness, but delivering anxiety and systemic dissatisfaction.'[16]

In the context of economics, the left hemisphere loves GDP as a reductionist, easily calculable proxy for human wellbeing, while the right hemisphere knows that many dimensions of human wellbeing can't be encapsulated in a number, especially a number that celebrates the conversion of nature into consumer products.

The Inescapable Hall of Mirrors

It's not just that the left hemisphere is running amok – it's becoming a self-reinforcing spiral. The more powerfully the left hemisphere creates the world in its own image, the less influence the right hemisphere has, and because of the right hemisphere's rather retiring, un-self-confident nature, it doesn't fight back and assert itself.

It doesn't help that, increasingly, communication is mediated through technology that gives precedence to the written word. In *The Alphabet Effect*, author Robert Logan writes:

> 'A medium of communication is not merely a passive conduit for the transmission of information but rather an active force in creating new social patterns and new perceptual realities.'[17]

He sees the shift from the oral traditions to written language as having been a key factor in the suppression of the divine feminine:

> 'Writing was a gift eagerly accepted by the ancients. Unfortunately, hiding among the neat rows of carefully incised script was an unwelcome demon – misogyny.'

We could equally see the rise of writing as a driving force in the ascendance of the left hemisphere. Historically, much of human communication was non-verbal – gestures, body language, pheromones, touch and that ineffable sense of a person's energy that attracts or repels. When a person is speaking to us, we listen (well, some of us do) but we also gather a lot of information subconsciously by noticing their tone of voice and facial expression. Listening is a holistic experience.

Speaking is also a holistic, dual-hemisphere experience. We gesture with both our hands, we exercise the muscles on both sides of our face, we use our tongue and our vocal chords and our lungs. Even while speech is generated in the left hemisphere, the right hemisphere is also active in sensing how our words are being received in real-time interaction with our audience.

By contrast, reading and writing are predominantly left-brain activities. Rather than gesticulating with both hands, we pick up a pen in our dominant hand, usually the right. We write linearly across the page. The written word lacks the depth and nuance of tone of voice, facial expression or gesture – as you'll appreciate if you've ever had a joke fall flat or a sensitive message be misunderstood when communicated by text or email rather than in person. And yet our communications are becoming more and more abstracted, with in-person meetings increasingly rare. The role of the right brain in social interactions is being increasingly downgraded.

Leonard Shlain also sees correlation (and, he implies, causation) in the emergence of written language and what Riane Eisler would call a dominator culture. In *The Alphabet Versus the Goddess*, he asserts that, throughout the world, the arrival of written language coincided with the shift from egalitarian to hierarchical societies, in which certain segments of the population (women, slaves) were compelled to work for another segment, often a minority.

Combining this with Iain's concern that the story of the Western world is one of increasing left-hemisphere domination, we see a pattern emerging, a chain of causality that starts between our ears. From a balanced partnership of hemispheres, the left hemisphere has come to dominate the right – which seems almost inevitable, given that the left hemisphere likes to dominate, manipulate and control things. Having conquered the right hemisphere, the left has gone on to create a world in which the masculine is valued over the feminine, the brain over the body, the rich over the poor, the big over the small, the material over the immaterial, and humans over the natural world. With every new manifestation of its values in the external world, the left hemisphere becomes more and more assured of its own supremacy.

I sometimes think of the hemispheres as two lovers locked in the drama triangle of persecutor-victim-rescuer. The left hemisphere is barging around doing things in its distinctive reductionist, materialist way. It's not trying to be a bully – if it thinks of it at all, it simply regards the right hemisphere as either irrelevant or subservient. The right hemisphere goes into victim mode, becoming increasingly withdrawn and morose. Meanwhile, the lines of communication between them become ever more constricted as the corpus callosum gets thinner.

I find my imagination stepping into the role of the right hemisphere, trying to have that *we need to talk about our relationship* conversation with the left hemisphere. I manage to distract him for a moment from his left-brained busyness, constructing his world of machines and commerce and technology, and we sit down with a cup of tea. I try to say that we've lost the magic, that he no longer listens to me and that if this relationship is going to work out, we need to communicate better and show mutual respect.

But then he gets all clever with me, waging his formidable powers of reason against my tender feelings – and it doesn't help that he has the monopoly on language, while all I can do is sit and emote. He masterfully deploys his skills of coercive control, overwhelming my fragile self-esteem by sheer force of will, articulating all the ways in which he allows me freedom to be myself – don't we go to visit art galleries and music concerts and book readings, and those bits of nature he hasn't built on yet? I'm thinking *that's really not very much*, but then suddenly he's gone, slamming out of the door to conquer the world and mould it in his likeness. And I'm left nursing my cup of tea, feeling hopeless and melancholy, and generally right-hemispherical.

Chapter 7

Systems

> 'The world is a complex, interconnected, finite, ecological-social-psychological-economic system. We treat it as if it were not, as if it were divisible, separable, simple, and infinite. Our persistent, intractable global problems arise directly from this mismatch.'[1]
>
> Donella Meadows

'Roz needs to pick just one thing, and focus on it.'

Apparently a dear friend and fellow environmentalist said this about me. I wasn't there, but it was reported to me afterwards. From her perspective, I seemed to have a different cause every year – plastic pollution, climate change, habitat destruction – which made me look scattered, or even scatty.

But here's what was going on from my perspective: I couldn't pick just one thing.

First, there were too many things to pick from. Even a single issue like plastic pollution seemed to come in 100 different flavours – was I going to focus on plastic bag bans, drinking straws, biodegradable drinks bottles, litter picks or ocean clean-ups?

Second, it seemed to me that everything was connected to every other thing. Plastic pollution was connected to consumerism, which was connected to the drive for growth, which was connected to capitalism. Climate change more or less ditto, and habitat destruction likewise. In its turn, capitalism was connected to economics, politics, human wellbeing, inequality and scarcity.

So, because *everything* was connected, no problem could be solved in isolation. It was almost easier to change everything, than to change one thing.

Events, Patterns, Structures, Mental Models

In late 2016, a few months after The Stifling Hand of Control conference, a friend pointed me in the direction of the *Living Planet Report* from the World Wide Fund for Nature (WWF) and the Zoological Society of London (ZSL). Amidst the sophisticated graphics of the glossy report, it was a simple diagram that caught my eye. It showed a stylised iceberg alongside a cartoonish ship. The top, visible part of the iceberg was labelled 'Events'. Beneath the ocean surface lurked the bulk of the berg in three layers, labelled in descending order as 'Patterns', 'Systemic Structures' and 'Mental Models'. (It was unclear whether the ship had been included for purely artistic reasons, or if it was a covert warning that humanity is about to wreck itself on the submerged layers of the iceberg by ignoring the underlying causes of visible problems.)

A lightbulb went on over my head, a lightbulb that has continued to shed its light on the way I see the world. I immediately knew that this iceberg diagram was a key piece of the puzzle in my understanding of how change happens – or fails to.

I'd tried to be well-informed about what was happening in the world by reading news media, but my mental picture of the news-scape always seemed fragmented and confusing (hello, left-brained news reporting). I didn't understand why things happened the way they did, or more to the point, why people behaved the way they did, so I couldn't see how we could ever behave differently.

Now this simple image made me realise I'd been focused on the visible events in the daily news, but the really important stuff lay in the layers of the iceberg hidden beneath the surface. It was the patterns, systems and mental models that were responsible for the way the world worked – and where change had to happen. No matter how good our individual intentions, 'A bad system will beat a good person every time', to quote the American engineer, W. Edwards Deming.

Too often, we get distracted by the tip of the iceberg, the events that occupy most of the daily news reports – a war, the latest murder, a politician's gaffe, celebrity gossip. The following dogmatic statement, commonly attributed to Eleanor Roosevelt, implies an unpleasant sense of superiority, but has also prompted me to choose more carefully what I pay attention to, because who wants to be called small-minded? I certainly don't.

> 'Great minds discuss ideas; average minds discuss events; small minds discuss people.'[2]

Leverage Points

The year after the WWF/ZSL report, an American friend sent me a copy of Donella Meadows' work on the leverage points in a system. Donella was the lead author of the Club of Rome-commissioned 1972 report, *The Limits to Growth*,[3] and identified the twelve most powerful leverage points in a system. The way she tells the story is that she was in a meeting about how to make the world better via the new trade agreements, the North America Free Trade Agreement and General Agreement on Tariffs and Trade, and the World Trade Organisation, and a sense of the utter wrongness of the premise started to simmer inside her.

> '"This is a HUGE NEW SYSTEM people are inventing!" I said to myself. "They haven't the SLIGHTEST IDEA how this complex structure will behave," myself said back to me. "It's almost certainly an example of cranking the system in the wrong direction – it's aimed at growth, growth at any price!! And the control measures these nice, liberal folks are talking about to combat it – small parameter adjustments, weak negative feedback loops – are PUNY!!!"'[4]

Almost before she realised what she was doing, she leaped up, marched over to a flipchart, and wrote a list of places to intervene in a system, in increasing order of effectiveness. The points she jotted down that day evolved into the following list:

12. Constants, parameters, numbers (such as subsidies, taxes, standards).
11. The sizes of buffers and other stabilizing stocks, relative to their flows.
10. The structure of material stocks and flows (such as transport networks, population age structures).

9. The lengths of delays, relative to the rate of system change.
8. The strength of negative feedback loops, relative to the impacts they are trying to correct against.
7. The gain around driving positive feedback loops.
6. The structure of information flows (who does and does not have access to information).
5. The rules of the system (such as incentives, punishments, constraints).
4. The power to add, change, evolve, or self-organize system structure.
3. The goals of the system.
2. The mindset or paradigm out of which the system – its goals, structure, rules, delays, parameters – arises.
1. The power to transcend paradigms.[5]

She rounds off the story of that moment of sudden insight:

> 'Everyone in the meeting blinked in surprise, including me. "That's brilliant!" someone breathed. "Huh?" said someone else. I realized that I had a lot of explaining to do.'[6]

And she continued to do a lot of explaining until her premature death in 2001, at the age of 59, from cerebral meningitis. I wish she was here to do yet more explaining, because we clearly haven't got the message yet. When we look at the list of leverage points, we seem to be mostly still dabbling around between numbers 12 and 6. Carbon taxes and fines for misbehaviour are too often seen as acceptable costs of doing business by the powerful multinationals that are the main environmental culprits. Polly Higgins' ecocide campaign to designate destruction of ecosystems as an international crime is getting closer to the top, round about line 5. But it's when we get to 4, 3, 2 and 1 that the leverage points get really exciting.

Richard Heinberg, writing for Ecowatch in 2017, credits *The Limits to Growth* for framing 'the modern human predicament in terms that revealed the deep linkages between environmental symptoms and the way human society operates', but laments that in the years since then the focus has narrowed to climate change in isolation, ignoring its systemic interrelationship with other ecological issues such as population, extinction, pollution and topsoil loss. He reflects:

> 'If climate change can be framed as an isolated problem for which there is a technological solution, the minds of economists and policy makers can continue to graze in familiar pastures.'[7]

The trouble with allowing policy makers to stay safely inside their comfort zone is that it simply won't work. This is where the left-brained approach of chopping reality up into little pieces lets us down in the most catastrophic of ways. Big systemic problems can't be solved little by little, battling one issue at a time in hopes of winning the war. Twenty years after Donella died, we're still mostly pulling on the puny little levers, because it takes a bold and brave leader to reach for the big levers of change.

What Would Donella Do?

Everything is connected. This is both a blessing, and a curse. The curse is that most problems turn out to be inextricably connected with a whole load of other problems and evolve even as we try to solve them. Our political, economic and social structures, like the problems they've generated, are interconnected and mutually reinforcing. And this is the blessing – that if we can find a leverage point that turns our vicious spiral into a virtuous one, when one thing changes, it could change everything.

But in a complex system, how do we figure out what that one thing is? And how do we predict what will happen when we change it?

In short, we can't. We can't be sure that we've identified the correct one thing, and we can be even less sure what will happen when we tweak it. But that's no excuse for remaining passive bystanders, drifting along in the status quo. What we do know is that superficial changes just aren't going to cut it.

Certainly, the human species as we know it will go extinct at some point, but when and how, and whether it is at our own hands or for reasons beyond our control, are still anybody's guess – as is the question of what we do in the meantime. Here are three broad scenarios:

1. **Humanity continues on its current trajectory**. Over the last few decades we've proved adept at researching and documenting the data that seem to point inexorably towards our demise, and at writing pretty resolutions full of nice words (Universal Declaration of Human Rights, Millennium Development Goals, Sustainable Development

Goals, the Paris Agreement, etc.). We have, however, proved less adept at taking corresponding action, so I am sceptical that we will make the radical course correction we need. But I expect a substantial number of humans will survive, and will adapt to a radically changed world. We may end up living on an impoverished planet where a strip mine sits alongside a shopping mall alongside a landfill, and nature has been entirely subjugated to serving humanity's needs. Maybe the ecomodernists and transhumanists turn out to be right, and we will be able to escape this miserable planet by moving to Mars, uploading our consciousnesses into a computer and augmenting our bodies until we are virtually immortal. This is not a future I relish. I'm really rather fond of this planet, and would much prefer that we learn to take better care of this one before we go screw up another one.

2. **Humanity takes radical action without a shift in consciousness.** We've already passed the tipping point on many of the critical planetary boundaries,[8] so it's likely that eventually we'll pay attention and make changes that may be sufficient to secure our survival. But without a fundamental shift in the way we think of our relationship with the Earth, any changes will be shallow rather than deep-rooted, and once the immediate crisis has passed we are likely to relax and recidivate to previous ways of being. The inevitable will have been postponed rather than permanently averted.
3. **Humanity takes radical action combined with a shift in consciousness.** This is the only path that is sustainable, in both the environmental and psychological senses of the word. We need to transcend our existing paradigm of domination and separation, and create a new narrative of radical interconnectedness in which we take responsibility for the wellbeing of all sentient creatures on Earth, both present and future, and then design our systems – political, environmental, social, technological, legal and economic – in alignment with that narrative.

We Can't Use Old Paradigm Thinking to Create the New Paradigm

Three-time Pulitzer Prize winner and free trade advocate Thomas Friedman, in a 2018 interview for *The Sustainian*, said provocatively:

> 'There is only one force bigger than Mother Nature, and that is Father Greed.'[9]

He was concerned, rightly, at our lack of progress on the United Nations' 2030 Global Goals, so was suggesting that politicians can spark massive investment in sustainability by leveraging corporate greed. I quoted the above remark in my blog and asked my readers:

> 'I'd love to know how that idea is landing with you right now. Does it seem like a sensible, pragmatic plan to incentivise change? Does it fill you with revulsion? Or somewhere in between? Myself, I'm somewhere in between, but leaning towards the revulsion end of the spectrum.'

First, I've been on the sharp end of Mother Nature for 520 days and nights on the ocean, and she's pretty damn forceful. And that was only the tiny little Earthly part of Mother Nature. She is, essentially, the entire cosmos, with its mind-boggling array of explosive supernovas, dazzling gamma-ray bursts, stars that can power themselves for billions of years, and oh, that little old Big Bang thingie that started it all. I reckon Mother Nature trumps human greed any day of the week, and twice on Sundays. But that is a pedantic point compared with my bigger objection.

There was something so desperately nihilistic about the statement, a strategy of last resort, attempting to use humanity's worst quality to achieve our redemption. Can billions of wrongs ever make a right?

I was reminded of a workshop I attended five or six years before I read Friedman's comment. It was called Values and Frames, and was run by a non-profit called Common Cause. As we arrived at the workshop venue in the morning, we were asked what we thought were the biggest challenges facing the world. Our answers were compiled on a flipchart – climate change, habitat destruction, deforestation, poverty, inequality, social injustice, gay rights, human trafficking, corruption ... The usual suspects.

Once we were seated, we were asked to look at a chart of fifty-seven values, and put marks next to the ones that we thought would be most useful, and least useful, in addressing those challenges. These values had been identified by psychologist Shalom Schwartz, whose Theory of Basic Human Values recognises ten universal forces that motivate human behaviour: self-direction, stimulation, hedonism, achievement, power, security, conformity, tradition, benevolence, universalism and spirituality.

The fifty-seven values fall within these ten broader categories. (See the excellent *Common Cause Handbook* for more information.[10])

The consensus of the workshop participants, which is borne out by research, was that the most constructive values were the intrinsic values, broadly classed as benevolence and universalism. The least useful were extrinsic values such as hedonism, achievement and power.

Further, we were warned about the danger of appealing to extrinsic values in order to motivate 'good' behaviours. Rewards, incentives and bribes were considered to be at best successful in the short term only (as parents of small children will testify), and at worst to condition people to expect rewards for actions that should be intrinsically motivated.

So at this point it looks as if Thomas Friedman is taking us into dangerous territory if we try to get people to do the right things for the wrong reasons. But are we assuming that all people are the same? Could it be that a more nuanced approach is possible?

In *How to Win Campaigns* (a refreshingly self-explanatory title), Chris Rose references the Cultural Dynamics analysis that tells us there are three kinds of people in the world (or at least in the UK):

> **Settlers** (31%). Motivated by resources and by fear of perceived threats. They tend to be older, socially conservative and security conscious. They are often pessimistic about the future, and are driven by immediate, local issues affecting them and their family.
>
> **Prospectors** (37%). Driven by the esteem of others. They are motivated by success, status and recognition. They are usually younger and more optimistic. They are often conscious of fashion or image, and tend to be swing voters.
>
> **Pioneers** (32%). Motivated by self-actualisation. Their views are governed by considerations of collectivism and fairness. In their personal lives they are ambitious, but seek internal fulfilment rather than the esteem of others.[11]

I identify as a Pioneer. If you're reading this book, I'm guessing you do too.

To use Daniel Kahneman's terminology, you could frame this as being happy *in* your life (Prospector or Settler) versus being happy *about* your life (Pioneer).[12] So we can imagine that a Pioneer would not, and should not, be moved by external incentives. But a Prospector might, if it made them look good. And a Settler would be influenced by issues affecting their family and community more than by anything Thomas Friedman might have to say. So arguably there is a role for external motivations, even if they only work for slightly over a third of the population.

I wonder if Friedman would respond by saying I'm creating a straw man argument, that he was talking about appealing to the greed of *corporations*, not individuals, which are not the same thing (no matter what US law may say on the matter), and because corporations have a profit motive, they are and *should be* motivated by greed.

But then are we in danger of reversing the encouraging trend towards companies doing well by doing good? Can't companies be intrinsically motivated, just as individuals are?

Swimming Upstream

What would Donella make of all this?

Photographs of her show a round-faced woman with a pudding-bowl haircut. In some she is relaxing, wearing wellies and leaning casually against a tree, or sitting on a stump, propped against the wall of a barn while sheep peacefully munch hay in the background. She was a research fellow at the Massachusetts Institute of Technology (MIT) under Jay Forrester, the inventor of system dynamics, whom she credits as a major inspiration.

Donella's husband, Dennis, was one of the three other authors of the *Limits to Growth* report. Together they founded the International Network of Resource Information Centres, informally known as the Balaton Group after the region of Hungary where their annual meetings still take place. She had a keen interest in sustainability and environmentalism, and was sceptical of the aspiration to never-ending growth. She and Tim Jackson would have got on well.

Yet growth is one of our most prevalent paradigms. Our addiction to growth is the mental model at the bottom-most level of the iceberg, a paradigm that we need to transcend if we are to have any hope of a sustainable future. No matter which way you defend, explain and justify it, infinite growth on a finite planet is a physical impossibility – and even if we could find a way to do it sustainably, the current growth model exacerbates social and economic inequalities.

However, society as a whole does not welcome perspectives that run counter to the mainstream narrative. Evolutionary biology has selected for humans who are happy to go along with the majority, because in earlier times, ostracisation from the tribe was as good as a death sentence. There is a theory that such rebellious souls did historically exist, either as individuals or as small communities, in the forests and mountains outside the villages where more conventional folks lived. They were known as savages, and given my last name, were quite likely my ancestors. If so, this explains a lot.

I am reminded of the cautionary tale of *The Emperor's New Clothes*, by Hans Christian Andersen, in which two swindlers resolve to get their revenge on an emperor who spends extravagantly on his wardrobe while his citizens suffer. They tell him that they will make him the finest robes from a very special cloth that can only be seen by the intelligent and wise, but is invisible to the foolish or incompetent. They sit at their loom, pretending to weave cloth, and all the courtiers who come to check on their progress act as if they can see the cloth, not wanting to be taken for fools. The great day arrives when the garments are supposedly ready, and the swindlers mime the dressing of the Emperor, who also pretends that he can see his new suit. As he parades through the streets of the city, the people all see that he is naked, but nobody wants to admit it for fear they are the only one. It is finally a child who blurts out, 'But he's not wearing anything!'

How many of us are like the crowd, aware that something is amiss, but because everybody else seems to be okay with it, we go along with the lie? I know that I was exactly like that when I was working in the corporate world. It seemed to me to be bordering on the unethical that we junior management consultants, still wet behind the ears from university and with minimal experience of the real world, were being hired out at extortionate rates to help 'improve' businesses. But I felt like I was the only person thinking such heretical thoughts. My peers and managers seemed to find it perfectly normal and acceptable, so I assumed it must be something wrong with me, rather than with the system.

This brings us back to the bystander effect. We might feel something is wrong, but if nobody else is reacting, we don't either. We push it down, not wanting to rock the boat or upset the applecart, not wanting to be cast out of the tribe. We doubt the evidence of our own senses until somebody like the child in the Emperor's crowd, like Greta Thunberg or Edward Snowden, comes along and names the thing that had been niggling at the back of our minds, and suddenly everybody admits what they had known all along but were too afraid to admit – that the Emperor of twenty-first-century civilisation is naked.

Looking at the Backs of our Heads

Trying to escape the matrix of the dominant paradigm can feel a lot like trying to look at the back of our own heads. There are clear-sighted thinkers who are signposting what this global matrix is made of and what keeps us trapped here, maybe offering a red pill to get out of it – yet even with all this information it can be incredibly hard to see things in ways we've never seen them before, especially when we live our day to day reality inside mutually reinforcing superstructures that, either by design or by default, protect themselves even while they hide in plain sight.

From cradle to grave, we're embedded in systems that don't serve us. To get hyperbolic for a moment, we have an education system that teaches us compliance, to prepare us for jobs that suck our souls, in corporations that destroy the planet and contribute to a metric of success (GDP) that values little that we value and plenty of things that we don't. We have a capitalist system that enriches organisations we don't trust, who fund politicians we didn't vote for, and the richer the corporations get, the more political power they buy. We have a consolidated, corporately owned, politically biased, fear-based media that keeps us superficially informed and deeply distracted, while ridiculing or deplatforming perspectives that run counter to the interests of its owners.

Worse, we are complicit in all of this. We have cognitive biases that focus our narrow spotlight of attention on what we already know and confirm what we expect to see, blinding us to alternative viewpoints and even blinding us to our own blind spots. As Boston College Professor Emeritus Bill Torbert said, 'If you do not understand your role in the problem, it is difficult to be part of the solution.'[13]

And all of these things – the economics, the politics, the media and our biases – are embedded within a dynamic of domination that starts inside our own skulls with the left hemisphere dominating the right, cutting us off from the wisdom that might just help us escape this quagmire. This dynamic radiates all the way out to the furthest corners of our civilisation, ensnaring just about everybody and everything in a dog-eat-dog world, and when we're not busy trying to eat or avoid being eaten, we're chasing our own tails, sensing that there is something terribly wrong, but trapped in a loop that keeps us running in circles, getting increasingly dizzy but apparently powerless to stop ourselves.

How do we extricate ourselves from this insanity? It often feels as if the more we struggle against the status quo, the deeper we sink into it. To mix my metaphors, can we ever find a sword sharp enough to slice through this maddening Gordian knot?

This may seem hopeless. It's about to get worse.

But hang on in there – after we've sized up the challenges that we're facing, I will offer a way out of the morass. It won't be easy, but it's not impossible. Just stay with me while we briefly consider the end of the world as we know it.

Chapter 8

Collapse

> 'Birth is painful and delightful. Death is painful and delightful. Everything that ends is also the beginning of something else. Pain is not a punishment; pleasure is not a reward.'[1]
>
> Pema Chödrön

Cassandras have shown up in my life at various times over the years. In the early 1990s there was a woman who caught me at a vulnerable and sleep-deprived moment on a ferry in Greece, and convinced me the world would end at the turn of the millennium. I overcame my gullibility once I'd caught up on some sleep, but her words left me with a long-lingering shadow of uncertainty over the future.

Then, in early 2004, about six months after my return from Peru, I spent several weeks alone at a cottage outside Sligo in Ireland for a self-imposed retreat. My rucksack contained a few clothes and a whole load of books, and I dived deep into various topics – the environment, life after death, philosophy, spirituality, indigenous culture and apocalypse. I do so enjoy a bit of light reading.

The books included *Coming Earth Changes*, based on the predictions of Edgar Cayce, the early twentieth-century clairvoyant known as the Sleeping Prophet due to his style of channelling while in a trance state. As the book title implies, Cayce predicted difficult times ahead. I found it particularly interesting that he drew a causal link between global devastation and human spiritual consciousness:

> 'Tendencies in the hearts and souls of men are such that these [upheavals] may be brought about. For, as indicated through these channels oft, it is not the world, the Earth, the environs about it nor the planetary influences, not the associations or activities, that rule man. Rather does man – by his compliance with divine law – bring order out of chaos; or, by his disregard of the associations and laws of divine influence, bring chaos and destructive forces into his experience.'[2]

He went on to foresee a dramatic movement of the Earth's magnetic poles:

> 'The earth will be broken up in the western portion of America. The greater portion of Japan must go into the sea. The upper portion of Europe will be changed as in the twinkling of an eye. Land will appear off the east coast of America. There will be the upheavals in the Arctic and in the Antarctic that will make for the eruption of volcanos in the Torrid areas, and there will be shifting then of the poles – so that where there has been those of a frigid or the semi-tropical will become the more tropical, and moss and fern will grow.'[3]

That could really spoil your day, especially if you live in California or Tokyo.

This might sound far-fetched – indeed many of Cayce's prophecies have failed to come true – but its impact on me was profound, especially because, inspired by my experience on the retreating glacier in Peru, I was also reading about catastrophic climate change, biodiversity loss and other forms of ecological collapse. These two strands – dire prophecies and environmental angst – interwove, each intensifying the other.

Also on my reading list in Ireland was *The Hopi Survival Kit*,[4] describing the prophecies of the Hopi tribe of North America. This choice was an indirect result of my Peru adventures; in Cusco I'd attended an evening event co-hosted by indigenous Peruvians and Navajo people participating in a wisdom exchange. It had motivated me to learn more about Native American culture. The traditional Hopi believe that we are on the cusp of transformation; that we are currently in the Fourth World, and on the threshold of the Fifth. Each of the three previous worlds have seen the destruction of humanity through war and self-destructive practices. When one world comes to an end, a few humans are permitted to escape the carnage and ascend to the next world.

Ever since the Second World War, the Hopi have been sending delegates to the United Nations to warn the world that their prophecies are coming true; that humanity is heading for a huge but turbulent transformation, and the state of our consciousness will be key. In drawing connections between our behaviour and our destiny, Cayce and the Hopi seem to be singing from the same eschatological songsheet.

Alone in my cottage in Ireland, without the company of sceptical friends to tell me to stop being so daft and doom-laden, these ideas rooted deeply. The impact of the ideas receded once I got home and back into the social whirl, but, as with the words of the crazy lady on the Greek ferry, never completely disappeared.

Prophets of doom have presumably existed for as long as humans have had the vocabulary to talk about the end of the world as we know it. I would guess that those who ridicule such prophets have existed just as long. Life has not got any easier for doom-mongers since the boy cried wolf over the Y2K bug at the turn of the millennium, and again over 2012 and the Mayan calendar. So it's easy to poke fun, but remember that in Aesop's fable, the wolf did eventually show up. One day, sooner or hopefully very much later when the sun extinguishes life on Earth, the prophets of doom *will* be right.

For many years, I took the prospect of environmental catastrophe as a call to action. I fiercely resisted any expressions of defeatism or resignation. In 2016, I spoke at Secretary of State John Kerry's Our Oceans Conference in Washington DC, between John Kerry's opening remarks and then-President Barack Obama's speech; I expressed the need for hope, concluding with a quote from the novelist Raymond Williams:

> 'To be truly radical is to make hope possible, rather than despair convincing.'[5]

But over the last few years, the idea of impending apocalypse, or at least the collapse of civilisation, has been creeping ever closer. It lurks in the shadows of my mind, like my own mortality – something inevitable, I just don't know how or when.

I'm not alone. Some fine minds have also arrived at the conclusion that our relatively blessed period of human history is drawing to a close. Here's what I've gleaned from their analyses.

(This might be a good moment to pour yourself a stiff drink, although definitely not before driving, and preferably not before breakfast.)

What's Got Us to Here Won't Get Us to There

Politicians, business leaders and technologists tend to talk as if human progress will continue in a smooth trajectory, and the past will be a good guide to the future. There is even a name for this particular cognitive fallacy – *linear projection*, defined by the environmental social scientist Adam Dorr as the error of 'presuming that future change will be a simple and steady extension of past trends'.[6] (And if you think that, even if the collapse does happen, it won't affect you, beware of the fallacy of *future personal exemption* – the error of being overly optimistic about your individual future despite the prognosis for the collective, like the smoker who believes the health warning applies to everybody but them.)

There is one sense in which the past *is* a good guide to the future – but it requires us to look deeper into the past than a single human lifetime, which is why we tend not to be very good at it. Remember the *availability heuristic* – things that don't come easily to mind tend to get elbowed aside by things that do.

It's important to my case that I convince you of the likelihood of imminent collapse, so I'm going to lay it on thick.

(Feel free to pour yourself a top-up before we start.)

In *Immoderate Greatness*,[7] William Ophuls provides a compelling synopsis, based on the historical record, of six major factors that propel civilisations toward breakdown. His blunt assessment of the situation is that 'civilization is effectively hardwired for self-destruction'. If you can find ways to refute his arguments, please tell me, because I'd like to believe it's not true – but his case is quite convincing.

So let's dive into his cheery hypothesis and the six factors, many of which conveniently start with E: ecological exhaustion, exponential growth, expedited entropy and excessive complexity, and then the non-Es of moral decay and practical failure.

1. Ecological Exhaustion

As a civilisation expands, so does its ecological footprint. To date, every civilisation has centred around a great city (Rome, Athens, Istanbul, etc.) and cities tend to: a) have a higher per capita footprint than the average for the rest of the nation, and b) shield from their inhabitants the extent of the ecological impacts (soil degradation, deforestation, declining aquifers, etc.) that happen outside the city limits. Part of the problem is the time lag before the ecological exhaustion becomes evident. The benefits are immediate. The costs are delayed. By the time they become apparent, the civilisation is already inexorably committed to overshoot.

Global warming is a classic case of time lag. Estimates vary, but according to climate scientists the current run of record-breaking average annual temperatures reflects the emissions of between ten and forty years ago. The ocean has been absorbing much of the excess carbon dioxide in the atmosphere, becoming more acidic in the process. Massive as the oceans are, they are now reaching saturation point.[8] They can't continue to flatter our emissions figures indefinitely. Conversely, even if we drastically cut emissions right now, there will be a time lag before we see the effect. We're living in a fool's paradise, and are probably already committed to several degrees of warming.[9]

Once one climatic domino falls, a number of further environmental dominoes are likely to fall in rapid succession. As sea ice melts in the Arctic, the meltwater pouring into the North Atlantic reduces the salinity of the ocean, which in turn disrupts the meridional overturning circulation, the technical name for the oceanic conveyor belt that weaves its way throughout all the world's oceans. This conveyor belt is powered by the density differential between saltwater, which sinks because it is denser, and fresh water, which rises because it is less dense. For several years now we've already been seeing the circulation system weakening as the North Atlantic becomes less salty with the influx of fresh water from the melting Arctic ice cap.

Because the Earth's oceans are all connected, the failure of the Atlantic thermohaline circulation also shuts down the Kuroshio, Leeuwin and East Australian Currents. These changes lead to hotter summers and colder winters in Europe, more violent storms on all the Atlantic seaboards, and an intensification of El Niño events in the Pacific. The Northern Hemisphere jet stream changes its behaviour, getting stuck in position for longer periods, leading to prolonged heatwaves, droughts and floods that exacerbate the already dramatic effects of the ocean circulation shutdown. A Great Climataclysm is eminently foreseeable, and yet we are doing very little to prepare.

Johan Rockström and his colleagues at the Stockholm Resilience Centre identified nine processes that regulate the stability and resilience of our global ecosystem, which are climate change, ocean acidification, ozone depletion, phosphorus and nitrogen cycles, biodiversity loss, freshwater use, land-system change, aerosol loading and chemical pollution.

I won't go into the specifics of what these phrases mean, but suffice it to say that besides climate change, we are also in overshoot on biodiversity, land-system change, and nitrogen and phosphorus. And our population and consumption patterns are still expanding. We're not just committed to overshoot – we are in it already. Revelations like these make me picture a Pac-Man gone mad, eating away the ground from under its own feet.

Another part of the problem is the *law of the minimum*, meaning that the variable that is in the shortest supply becomes the limiting factor. It doesn't help a civilisation much if it has bountiful building materials, or even food, if it doesn't have enough water. If the population is already at the full carrying capacity of its environment, and then that capacity is reduced over a short space of time – rising oceans engulf coastal areas, desertification destroys arable land, ocean ecosystems collapse due to acidification, and/or soil degradation reduces crop production – then we're really up the creek.

In many ways, it's not our fault. It's in the nature of every species. Bacteria in a Petri dish will use all the resources at their disposal until they have consumed all the nutrients and drown in their own waste. Just like bacteria, humans seem hell-bent on expanding as much as they can given the prevailing conditions, until further expansion becomes physically impossible. We would, of course, like to think that we are more intelligent than bacteria in this regard, but we have yet to prove it.

Oops (#1).

2. Exponential Growth

As a species, we're not good at comprehending non-linear growth. Think of the bacteria in the Petri dish again. If they double in number every day, they're probably still feeling pretty mellow when their dish is one-eighth full. 'Hey folks, loads of space – let's stretch out and relax!' The next day their dish is still only one-quarter full. Still roomy. The next day it's half full. Still no problem. The day after that …

Our population growth might be slower than bacteria, but it's still happening *really* fast. To once again use my lifetime as the benchmark, when I was born in 1967 the population was about 3.5 billion, up from about 2 billion in 1929 when my father was born. By the time I turned 44 in 2011, the population had doubled to 7 billion. I'm now 54, and we're closing in on 8 billion.

To place this in the context of deep time, I'll borrow the image used by palaeontologist Louise Leakey. We're going to use a toilet roll, not for its usual purpose, but to represent the history of planet Earth. If you lay out an average loo roll, it's 400 sheets long. The dinosaurs appear on the 382nd sheet, the 19th sheet from the end, and go extinct on the 5th sheet from the end, at which point the mammals start to emerge. So dinosaurs had fourteen sheets of historical time, mammals have five so far. Homo sapiens? We come into being in the last 200,000 years of Earth's history, which equates to *the last millimetre on the last sheet* of the roll. And the last fifty years, in which we have changed the world so much – you wouldn't even be able to see it. It's 250 nanometres, or 1/400 of a human hair.

Our rate of population growth is now slowing, largely due to the education of girls and the availability of contraception. Estimates of the human population in 2100 vary from 9.4 billion to 10.9 billion.[10] This, of course, assumes other things being equal (*linear projection* again), which they rarely are.

This might be a good time to mention another way of looking at our overshoot situation. Earth Overshoot Day is a metric calculated by the Global Footprint Network.[11] It looks at our current consumption of the Earth's resources, and computes the day of the year on which we go into overshoot, meaning we have used up all the resources the planet can regenerate in a year. It's calculated by dividing the planet's biocapacity (the amount of ecological resources Earth is able to generate that year), by humanity's Ecological Footprint (humanity's demand for resources in that year), and multiplying by 365, the number of days in a year. For a civilisation to be sustainable in the long run, it needs to be able to live on its ecological *interest* without digging into the *capital*.

In 1970, Earth Overshoot Day was 30th December,[12] meaning that we were only marginally exceeding the carrying capacity of the Earth. We would run out of our theoretical pool of global resources one day before the end of the year. By 1990, Earth Overshoot Day was 10th October. In 2010, it was 6th August. In 2022, it was 28th July. In other words, we would need 1.7 Earths to support the current number of humans at our current rate of consumption.

Obviously, some countries consume more than their fair share, while other countries consume considerably less. In 2022, the UK is projected to overshoot on 19th May (corresponding to 2.6 Earths), with the US and

Canada in a dead heat on 13th March (5.1 Earths). Only 50 out of the 189 countries listed don't overshoot and, as you would imagine, there is a strong correlation with poverty.

Ecomodernists may believe we will always find new ways to squeeze more resources out of our planet. I refer them back to the IPAT equation, and Paul Gilding's assertion that we have *never* managed to develop technology fast enough to keep pace with our increasing consumption.

In short, we are borrowing from future generations to subsidise today's lifestyles. This is unlikely to earn us the gratitude of our descendants.

Oops (#2).

3. Expedited Entropy

The Second Law of Thermodynamics tells us that systems tend towards chaos and disorder. In practical terms, this means that important resources like fossil fuels, precious metals and agricultural soil tend to degrade, requiring greater effort and increasing cost to yield the same results.

William Ophuls takes agriculture as an example.

> 'Agricultural production is the foundation of civilized life. But the word production is a misnomer, for what humans actually do is mine the topsoil. Virgin soil is a complex ecosystem developed over millennia that contains a myriad of chemical elements and biological beings within a very specific physical structure. Humanity breaks into this ecological climax to profit from the rich store of energy that it contains.'

Modern farming practices tend to emphasise short-term profits over long-term sustainability, so the soil is eroded, compacted, leached and otherwise degraded. Modern sewerage systems don't return the nutrients to the soil – instead they are flushed, treated and discharged into rivers and oceans, never to return.

Similar story with oil, minerals and metals. Used to be you could stick a nodding donkey anywhere in Texas and have a chance of striking oil. But all the easy stuff is gone, so now we're having to resort to difficult, dirty and costly sources of oil, like tar sands and fracking.

We may be getting more efficient, but we're still digging (literally) into our natural capital. Efficiency may delay the day we run out, but not forever.

Oops (#3).

4. Excessive Complexity

Civilisations tend towards complexity. Unless you, dear reader, are very young, you have probably observed this in your own lifetime. If you feel that life has got more complicated – more technology, more forms, more structures, more information, more *stuff* – you'd be right. Increasing sophistication leads to problems as technologies, structures and processes multiply and start to interact with each other. Attempts to solve the problems only add to the complexity until they amalgamate into a *problematique* of mutually aggravating feedback loops.

Thomas Homer-Dixon calls this the *ingenuity gap*. The accumulating problems outstrip human ability to innovate, until the chasm between the demand for ingenuity and the supply of it can no longer be bridged. It's rather like the T in IPAT again – we can hope that we'll be smart enough to outsmart our problems, but we never have been so far.

It was complexity (and greed) that was largely to blame for the 2008 financial crisis. Robert Peston writes in *How Do We Fix This Mess?* about his insight as he was touring the trading floor of Morgan Stanley, a major investment bank. He realised that if he didn't understand what was happening on the trading floor, then in all likelihood neither did the directors. The traders were, in effect, unaccountable for their extreme risk-taking.[13]

We can get carried away with our own cleverness, building a wobbling house of cards, layering complexity upon complexity until the whole structure becomes so precarious that it takes just a puff of wind to bring it tumbling down.

It seems we learned little from the 2008 crash. We keep trying to fix things, slapping successive plasters – legal, financial, structural – onto the problems, which only further adds to the complexity. We end up in a hopeless arms race, kicking the can of disaster down the road, but never actually taking the time to step back, simplify and consolidate the basics.

And so we end up in a *complex adaptive system*, in which there are so many variables, so deeply connected, that everything we do creates consequences, and predicting those consequences is nigh on impossible.

Oops (#4).

5. Moral Decay

In his 1976 book, *The Fate of Empires and Search for Survival*,[14] General Sir John Glubb examined the life cycles of eight empires since 859 BCE and concluded that each empire passed through the following phases:

> The Age of Pioneers: expansion of territory.
> The Age of Conquests: more expansion, not always peaceably.
> The Age of Commerce: wealth is created through trade and innovation.
> The Age of Affluence: all appears to be well, but the seeds of destruction are being sown.
> The Age of Intellect: the acquired affluence enables people to pursue the life of the mind. Academic institutions may produce sceptical intellectuals (and snarky journalists) who start to question the dominant narratives of the empire, undermining its authority.
> The Age of Decadence: people pursue happiness through excessive consumption, which in actuality makes them less happy. The powers that be create diversions for the populace, from gladiator fights to Facebook and Instagram, while people indulge in addiction and debauchery. The values and discipline that enabled the creation of the empire are eroded.
> The Age of Decline and Collapse: inequality grows (think Occupy and the 1 per cent), increasing numbers are excluded from meaningful work and the means to fulfil their potential (Gallup currently reports 80 per cent employee disengagement from surveys across more than 160 countries). Discontent leads to disruption and the empire collapses.

Glubb points to the heroes of an empire as a key indicator of where it is in its life cycle. During the early phases, pioneers and warriors are lauded. Then come the entrepreneurs and merchants. Once celebrities such as film stars, musicians and athletes become the main focus of popular attention, the empire is in trouble. Once the Kardashians dominate the tabloids, it's well and truly screwed. (I made that last bit up, but you get the point.)

William Ophuls describes the terminal phase in words that might resonate:

> 'Frivolity, aestheticism, hedonism, cynicism, pessimism, narcissism, consumerism, materialism, nihilism, fatalism, fanaticism, and other negative attributes, attitudes, and behaviours suffuse the population. Politics is increasingly corrupt, life increasingly unjust. A cabal of insiders accrues

wealth and power at the expense of the citizenry, fostering a fatal opposition of interests between haves and have-nots. Mental and physical illness proliferates.'[15]

Oops (#5).

6. Practical Failure

In his 2003 TED Talk,[16] Jared Diamond, author of *Collapse*,[17] lists three reasons societies fail to take action when collapse seems obvious and imminent. The first is the conflict of interest between the short-term goals of the ruling elite and the long-term interests of the society as a whole, especially if the elite are able to insulate themselves from the impacts. There is friction between short-term political election cycles and the long-term planning required to build resilience.

We read about wealthy people investing in escape hatches such as remote ranches, homes in gated communities, cryogenic preservation[18] and tickets to Mars.[19] I recall a conversation with a privileged woman at the Explorers Club of New York, in around 2007, when she told me about an off-grid property that she and some friends were buying in Idaho (I think, or some such sparsely populated state) so that when the excrement hits the rotating cooling device, they will be just fine. She invited me to buy a share. I declined.

The fundamental problem is this: the people with the power to change the system are the ones who have benefited from the system being exactly as it is. They have nothing to gain by changing it, and everything to lose. Rather than doing what is best for the population, they do what is best for themselves. As J.K. Galbraith said, 'People of privilege will always risk their complete destruction rather than surrender any material part of their advantage.'[20]

And so the status quo remains firmly in place, while civilisation goes to hell in a handbasket.

Oops (#6).

Our Tragic Flaw

So are we totally up the proverbial creek without a paddle? The fundamental problem is that humans are not good at recognising when enough is enough. As William Ophuls puts it:

> 'Those afflicted by hubris become the agents of their own destruction. Like a tragic hero, a civilization comes to a ruinous end due to intrinsic flaws that are the shadow side of its very virtues.'[21]

In other words, by the time the participants in the civilisation notice what's going on, it's already too late. We fall asleep at the wheel, assuming that what has worked in the past will continue to serve us in the future. Because we haven't consciously changed anything, but rather, drifted along with the ambient rate of change, we don't realise that we have fundamentally altered the terms of engagement with our ecosphere. In our culture of more, more, more, we forget about Aristotle's Golden Mean: that too much of a good thing becomes a bad thing. On the surface, all looks well until it suddenly goes from hunky-dory to Humpty Dumpty, and all the king's horses and all the king's men can't put our world back together again.

Based on his analysis of eight empires over the last 3,000 years (Assyria, Persia, Greece, Rome, Arab, Marmeluke, Ottoman, Spain, Romanov Russia and Britain), Sir John Glubb concludes that each civilisation spanned approximately 250 years, or ten generations. Every civilisation has risen, and has fallen. Likewise, every civilisation has believed itself to be indomitable, until it wasn't. The cycle is startlingly predictable.

(The US celebrates the 250th anniversary of its founding in 2026. Just saying. But writing from the relic of an empire that, according to Sir John Glubb, ended in 1950, I can assure you it's not so bad. We Brits still muddle along okay.)

But if this cycle is so predictable, surely we can extricate ourselves from the tyranny of history fated to repeat itself ... Right? Now that we know the dangers, we can avoid them. Okay, so no previous civilisation has ever managed to avert disaster, but we're the most developed society that has ever walked the Earth. If any generation has the ability to save itself ... don't we?

William Ophuls gives this hope short shrift:

> 'We could possibly do a better job of managing or arresting decline if not for one final, fatal factor. Human beings are barely evolved primates driven by greed, fear, and other powerful emotions ... In addition, humans are only partly rational, so they also suffer multiple mental aberrations – delusions, compulsions, manias, idées fixes, and the like.'[22]

This is where our cognitive errors well and truly come back to bite us in the rear. As we've already seen, we have well-catalogued lists of fallacies, and yet we all – psychologists included – still put the *fall* into *fall*acy with depressing inevitability. Escaping collapse is about as likely as escaping our cognitive biases, which is not very likely at all. Sadly, G.I. Joe gets the better of us every time.

As if that isn't enough, remember the reductionist, over-confident left hemisphere's exaggerated belief in its own abilities? The British historian Arnold Toynbee, in his twelve-volume *A Study of History*,[23] argues that successful empires expand by developing strong structures for problem resolution that they then try to apply doggedly to *all* problems. Their eventual collapse is not caused by losing control over the environment, or attacks from outside, but from their own stubborn refusal to change with the times. As Abraham Maslow was the first to say, in 1966: when all you have is a hammer, everything looks like a nail.

Once a civilisation starts to fall, the rigid thinking and self-satisfied complacency of the left hemisphere blind it to the inadequacy of its beloved structures and processes to handle the new reality. Despite growing evidence to the contrary, it becomes stuck in the mindset of *this has always worked before so it will surely work again*, and this attitude only exacerbates the death spiral.

How's that drink going?

This might seem like depressing stuff, hopeless even. According to Glubb, Ophuls and Toynbee, the karmic cycle of empires seems inescapable. Even if we confront our ecological crisis, overcome our cognitive biases, shift our paradigms, redesign our economics, overhaul democracy and burn the patriarchy to the ground, are we still fated to fail, cruising for a bruising, hurtling headlong towards an existential face-plant?

Yup, quite likely.

If you're feeling rather low right now, it's understandable. It's hard facing up to the massive challenges facing our civilisation, harder still to acknowledge the cognitive traps that have created and perpetuated these challenges, and harder yet to accept that even being aware of these cognitive traps won't help us escape them. You could be forgiven for thinking that we are well and truly scuppered.

But please don't throw this book aside in a fit of despondency. I don't want to leave you here in the midst of seemingly intractable problems coalescing into a perfect storm of existential threat.

Things do get better from here on in, I promise. We can yet chart a course from despair to optimism. I admit that, from the level of consciousness that created these problems, the future is not looking good.

But from a different level of consciousness, there is everything to play for.

PART THREE

ESCAPE

When we get out of the glass bottles of our own ego,
and when we escape like squirrels from turning in the cages of our personality
and get into the forest again,
we shall shiver with cold and fright
but things will happen to us
so that we don't know ourselves.

Cool, unlying life will rush in,
and passion will make our bodies taut with power,
we shall stamp our feet with new power
and old things will fall down,
we shall laugh, and institutions will curl up like burnt paper.

D.H. Lawrence, 'Escape', from
The Complete Poems of D.H. Lawrence (1957)

Chapter 9

Paradigms

> 'Edges are transitional places; they are also the best places from which to create something new. Ecologists call it the "edge effect": at the convergence, where contrasting ecological systems meet and mingle ... Those of us who live here [on the edge between land and sea] must be comfortable with storms and with change, for it is on these unsettled, unsettling edges that we will hear the Call which launches us on our journey. And though we can never quite be sure what that journey will involve, we know that new possibilities may be created only if we surrender to uncertainty.'[1]
>
> Sharon Blackie

It feels profoundly unnatural, at least to a landlubber like me, to set out from the safety and familiarity of dry land to row across an ocean. The departure from San Francisco in 2008, on the first leg of the Pacific crossing, was the one I found hardest.

The previous year I'd set out from Crescent City, an eight-hour drive north of San Francisco, to row the Pacific, but my attempt had ended in disaster and humiliation, airlifted reluctantly into the back of a US Coast Guard helicopter after a concerned wellwisher asked them to pluck me from stormy – but survivable – seas. It was now nine months later, and I was trying again with the intention of making landfall in Hawai'i, 2,400 miles away.

Since the failed attempt I'd spent most of my time in the San Francisco Bay Area, overseeing alterations to my boat and preparing for the next attempt. I'd had wonderful times there, and made many friends. Leaving land is hard enough, but even harder when leaving behind good people and happy memories.

The tide rushes out of the San Francisco Bay at a great rate of knots, the huge harbour decanting through the narrow gap between the headlands on either side of the iconic Golden Gate Bridge. It would be futile to try and fight the incoming current, so I was leaving with the ebbing tide. My weatherman and I had created a spreadsheet using a traffic light system, with red, yellow and green blocks ranking each hour's favourability in terms of wind direction, wind speed and tidal flow. Sadly, my beauty sleep was not deemed a relevant variable, and we settled on midnight on Sunday, 25th May as my best opportunity.

If you drive over the Golden Gate Bridge at night, heading north towards Marin, to your right you will see the bright lights of the city spangling the hillsides surrounding the bay. If you turn your head to look the other way, towards the ocean, you will see almost nothing but darkness, just a couple of navigation buoys and a lighthouse punctuating the night.

From the warm comfort of a car zipping across the bridge, you might spend a moment marvelling at the contrast between east and west, but soon you'll be on the other side, and your thoughts move on. From a rowboat, down in the choppy dark waters beneath the bridge, it's a very different perspective. With every oarstroke I put more distance between myself and all I had grown to know and love in the city, towards the Stygian darkness beyond the Golden Gate.

Liminality

As we've already considered, humans are wired to prefer the familiar. Evolutionary biology favoured those who stuck with what they knew, rather than the guy who wondered if this berry might be good, or the girl who wondered what might live in that cave. It takes a particular kind of courage, or stupidity, to do things that haven't been done before.

And yet, it's becoming increasingly evident that we're facing challenges we've never faced before, and if we hope to avoid the fate of previous civilisations, we need to avoid the mistake of applying old solutions to new problems.

Liminal ages are the times when we stand on the threshold of a new way of being, and liminal ages call for liminal thinking – the capacity to operate with imagination, creativity and insight at the threshold of what we know, as individuals and as a society. This isn't to underestimate the challenges, fears and losses of letting go of the known for the as yet unknown. It can be tough to let go of the old, even when it's starting to fail. We're not only rowing our boat into the darkness, but we don't have a map, and we don't know where we're trying to get to. Understandably, this prospect is enough to give most people the collywobbles.

Add to this that transformation is usually a messy business. As Deepak Chopra says, 'All great changes are preceded by chaos.' Chaos is an inevitable, unavoidable part of the cycle. Few of us in the industrialised world have experienced chaos at the societal level. Since 1945, war in our countries has been blessedly rare and, barring a few sudden leaps like the fall of the Berlin Wall, we've got used to change being so gradual and incremental that we barely notice it, and it mostly seems to be for the better. The next round of changes may not be so gentle.

Day and night, summer and winter, waxing and waning, ebbing and flowing ... This is the way of the natural world. In our human worldview, we tend to favour one side of the cycle over the other: we prefer the way up over the way down, life over death, summer over winter, growth over contraction. We label one aspect as good, the other as bad.

But Nature thrives on the dynamic interaction of yang and yin, hot and cold, light and dark, land and sea. Life and innovation flourish at the creative edge between order and chaos. Without change, there is no impetus for growth, no evolution.

Nature is periodically wracked by crisis – earthquakes, wildfires, volcanic eruptions, floods and droughts. We humans try to mitigate the effects of such disasters, but crises play an essential role in generating transformation. Nature knows when it needs a radical clearout of the old to make way for the new, to keep the system fresh and vital. It's only humans that get clingy and attached to things as they are.

When a caterpillar enters its chrysalis phase, what happens next is pretty alarming, not to mention disgusting. It digests itself with its own enzymes, creating a slimy soup from which the body of the butterfly begins to form. For a while it looks like complete annihilation, as if nothing viable can ever come out of this primordial chaos. And yet, it does – and not just something viable, but something stunningly beautiful. In this crucible of transformation, previously dormant imaginal discs emerge and multiply, fuelling the metamorphosis from the surrounding soup, and what eventually appears is not an upgraded caterpillar but a completely different creature – a butterfly. What once crawled now flies. What may once have been drab is now resplendent and iridescent. What used to munch its way stolidly through leaves now sups on nectar.

I don't suppose the caterpillar has much of an opinion about what it's about to go through. As far as we know, it's following its instincts, and is unlikely to be going through any existential angst over its impending butterflyhood. As humans, we're capable of a more complex cocktail of emotions. Even if we're excited about the prospect of becoming metaphorical butterflies, we may dread what we have to go through in order to get there. But the one sure thing in the world, more certain even than death and taxes, is change. We can resist it, or we can embrace it. Paradoxically, resistance is likely to end up being more painful.

The Hopi Elders issued a statement in 2000. I first read it in 2003, and there have been few days since when it hasn't crossed my mind. They offer this sage advice:

> 'There is a river flowing now very fast. It is so great and swift that there are those who will be afraid. They will try to hold on to the shore. They will feel they are being torn apart and will suffer greatly. Know the river has its destination. The elders say we must let go of the shore, push off into the middle of the river, keep our eyes open, and our heads above the water.'[2]

Collapse as Catalyst

It was February 1966 when American mother-of-five Barbara Marx Hubbard had a life-changing revelation while walking on a frosty hill near her Connecticut home. As she walked, she wondered what story might have equivalent power to the birth of Christ, a story that would change everything. Her mind went quiet, and a vision came to her:

> 'Suddenly, my mind's eye penetrated beyond the blue cocoon of Earth and lifted me up into the utter blackness of outer space. From there I witnessed the entire sweep of Earth's history, as though I were seeing a Technicolor movie. I saw the Earth as a living body, just as the astronauts did. It was alive.'[3]

She sensed the Earth struggling for breath, feeling the pain of the accumulated hunger, war, torture, disease and species extinction. In her mind's eye, she saw humanity sense the pain and, as one, stop in its tracks to pay attention to the Earth's suffering. Then she saw a flash of light.

> 'Empathy began to course through our planetary body and through all people ... I heard a tone, a vibration of resonance, connecting all of us in a moment of global coherence. .. I saw billions of us open our collective eyes and smile.'

Her belief was, now that humanity understands the process of evolution, we have the ability to shift from unconscious to conscious evolution, to determine our future direction as a co-evolutionary, co-creative species – if we simply have the right story to guide us.

She didn't see the collapse of civilisation as necessarily a bad thing, but, rather, a potential catalyst to take us to the next level if we're willing to admit that the human-made world in its present form isn't working. The failure of our current civilisation is bringing us to a fork in the road – it might jolt us into a higher order of consciousness ... or lead to our self-destruction.

Many traditional cosmologies also tell stories of failure, collapse and destruction as prerequisites to new and better ways of living, from Noah's ark, to the phoenix rising from the ashes, to the Aztec legend of the Five Suns. Although consciously we may resist change, the collective unconscious holds a deep archetypal resonance with the cycles and spirals of life.

Stability can turn to stagnation, becoming dissolute, decadent and ultimately moribund. Fortunately, it never happens. Change is always happening. The question is what kind of change we want to create.

Hard Times are the Greatest Teachers

When I look back over my life, I've definitely learned more from the bad times than the good. My first crossing, the Atlantic, was especially a sufferfest. I was pushed far beyond what I'd thought were my boundaries of pain, fear, boredom and frustration. There wasn't a day when I didn't wonder why this had ever seemed like a good idea, and why hadn't I let the call to adventure go find some other muggins to do its bidding instead of me.

At the same time, I knew I was learning a lot about who I was, who I could be and what I was capable of – but I thought I was learning these things *despite* all the challenges. It was only about ten years later that I realised that I had learned them *because of* the challenges. If everything had been easy and gone according to plan, I would only have acquired a fraction of the beautiful, hard-won wisdom that flooded in during those long and difficult days. I now call this the Gift of Hard Times – it's a blessing to know I can endure such difficulties, and find the inner resources to get me through.

Unfortunately, we don't discover these resources by sitting on the sofa waiting for them to show up. We have to actually do the thing we don't want to do. As Eleanor Roosevelt said (in one of her less snobby moods), 'A woman is like a tea bag – you can't tell how strong she is until you put her in hot water.' (Or cold seawater, in my case.)

So now I am less quick to label experiences as *good* or *bad*. What I used to think of as *good*, I now appreciate and am grateful for, while knowing they won't last forever. What I used to think of as *bad*, I now welcome as a learning opportunity (although usually my ego-mind has to throw a tantrum and sulk for a few days before I eventually remember that these times are a gift).

So, according to this belief system, discomfort is no bad thing. It's a cliché because it's true, that the magic happens outside your comfort zone. If personal growth and evolution is our purpose, then discomfort isn't the enemy – boredom and predictability are.

That's all very well, you might be thinking, but getting outside my comfort zone is, by definition, *un*comfortable. And that's not fun.

I agree – when we're going through strong emotions like grief, fear, failure, overwhelm or loneliness, we can often get caught up in the maelstrom. We identify with our emotions. We can't connect to any part of ourselves apart from the part that is hurting, which seems to occupy our entire consciousness. We feel completely immersed in negative feelings, consumed by them.

But it's a choice. Impossible as it may sound, there is always a part of us that is okay, a part that is still functioning. We can choose to identify with the fear, or we can choose to identify with curiosity, excitement and a desire to be strengthened by the experience. We can fight the change, futile as that will be, or we can allow the change to change us.

When I was on the ocean, struggling with pain, fear and frustration, I took great inspiration from the life of Viktor Frankl, the Austrian psychiatrist who spent three years in concentration camps, including Auschwitz. His father, mother, brother and wife were exterminated or died of disease. He endured hard labour while on starvation rations, wearing ragged clothing wholly inadequate for the bitterly cold winter conditions. Yet he held himself together, relying on what he termed psychological hygiene, as described in his book, *Man's Search for Meaning*:

> 'Mental health is based on a certain degree of tension, the tension between what one has already achieved and what one still ought to accomplish, or the gap between what one is and what one should become ... What man actually needs is not a tensionless state but rather the striving and struggling for a worthwhile goal, a freely chosen task. What he needs is not the discharge of tension at any cost but the call of a potential meaning waiting to be fulfilled by him.'[4]

So, in other words, we *need* to be stretched in order to thrive. We may think we want life to be a bed of roses, but we need a few thorns as well if we're not going to become weak and flabby. This is not to excuse the unalloyed evil of the Holocaust, but rather to revere Frankl for having the grace to find the gift amidst the horror.

This isn't me being macho and hardcore. Despite what you may think about a person who would voluntarily row solo across three oceans, and not for a moment to compare rowing an ocean with the ghastliness of Auschwitz, I can tell you honestly that taking on a challenge of that scale and magnitude was an aberration. I like an easy life as much as the next couch potato. Looking back, I marvel at how I did it, although I'm very grateful that I did because now I'm able to greet hard times with the knowledge that what won't kill me will make me stronger – or at least, hopefully, a little wiser.

Looking back over human history, it seems that change often happens – and possibly only happens – in response to a catastrophic breakdown of some sort. I had to hit a nadir of frustration and exasperation on the Atlantic before I could surrender my ego-mind and find a better way to cope. An addict often has to hit an absolute rock bottom of pain and self-loathing before they acknowledge they have a problem and turn their lives around.

After two world wars in four decades, humanity was so appalled by its atrocities that we resolved to never allow such things to happen again.

A contemporary of Viktor Frankl was Edith Eger, whose book, appropriately called *The Choice*, is equally inspirational. She experienced unimaginable horror, and yet she did not succumb to it. She used the horror to fuel her inner strength. The greater the devastation, the more she rose to the challenge, developing deep wells of wisdom and resilience. Her story reminds us that suffering presents us with a choice – to suffer, or to find freedom from suffering.

This isn't to say we shouldn't do our utmost to prevent war, environmental destruction or any of the other enormous challenges facing our world. Our endeavours as activists also unleash creativity, innovation and evolution. If activism is what calls you, then answering that call is what will unlock your greatest potential. It's all part of the evolutionary drama that we play out on the Earthly stage, all fuel to our evolutionary fire, and getting passionate about a cause can help us connect to enormous reserves of untapped potential.

It seems we may shortly be embarking on our greatest adventure yet as a species. There's a good chance we'll experience unprecedented challenges over the coming decades, be that the crumbling of civilisations, pole shifts, runaway climate change, the collapse of the oceans, antibiotic resistance, pandemics, solar flares or anything else from a long menu of existential threats.

We can try to ignore the crisis, or protect ourselves from it, but it's unlikely that even the richest among us will be able to fully insulate themselves from the impacts. A better strategy would be to welcome the gifts of growth that ride alongside the horsemen of the apocalypse, knowing that the greater the challenge, the greater the growth.

Living Life on the Edge

The liminal places are where all the cool stuff happens. We need to have one foot in each of the two worlds, the old and the new. We need one foot in the old order, because if we leave it entirely, we can't be a force for change. But we also need to have one foot in the emerging new order, from where we can see the strengths but also the weaknesses of the old system, and start to imagine a better world.

The problem for people who have both feet squarely in the old order is that they often can't see the need or potential for a new way of doing things. While Otto Scharmer says in the introduction to his book, *Leading from the Emerging Future*, that the old civilisation is collectively creating results that nobody wants, I don't think it's quite accurate to say that nobody wants them. There are clearly some bodies who do want those results, because the results suit them just fine, thank you very much.

The Franciscan priest Richard Rohr writes, 'Evil can hide in systems much more readily than in individuals.'[5] Many systems embed old and outdated moral codes. It has often been said of racial bias in the American (so-called) justice system that the system isn't broken; it was designed this way. We inherited systems from an era when there was less awareness and less respect for the rights of our fellow humans and of nature, but because we are so embedded in the systems, and *it has always been this way*, those of us who have nothing to complain about are often blind to their flaws. If the injustices are not part of our daily experience, too often we choose not to know what nobody can deny.

Conversely, having both feet in the new order will not generate impetus for change either. While we need some pioneers to role-model what's possible by living in an off-grid utopia, we also need people willing to bridge the old and the new, and to stretch out a hand to help others across the bridge.

So even though most of us can't yet see the future that wants to emerge, we can be fairly sure we're not going to see it from 5 miles inland. We need to step up to the edge of the cliff and gaze to the horizon.

Courage

This takes courage, which is something I've thought about a great deal. I don't identify as a particularly courageous person, so how did I find the inner resources necessary to row away from San Francisco, out into the pitch-dark Pacific?

Strangely, the disaster of the previous year's aborted attempt helped. I had no idea if this attempt would fare any better, but I knew I had to try. After the rescue, I'd taken a battering from the armchair critics, gleefully bashing out their messages of hate, scorn and condemnation from the comfort of their homes and the safety of their anonymity. To protect my hurt feelings, I told myself that my failure justified their own inadequacies and timidity: *haha, look at this daft woman – better to stay home than go taking risks!* But I still wanted to prove them wrong. I had faith in my boat and in myself, and I would let my actions speak louder than words.

Curiosity helped, too. After the misery of the Atlantic crossing, and the humiliation of the rescue, I was fairly sure something – or many things – would go wrong. I just didn't know what. But as the adventurers' saying goes, if you know how it ends, it's not an adventure. And I trusted that some things would go right – that I would see creatures I hadn't seen before, learn new things, discover new capacities. I was also curious about how well I could embody what I'd learned from the Atlantic; I'd worked hard to integrate the lessons, and the best test would be to get back on the ocean and see if I could do better this time around. Curiosity, for me, isn't just the point of adventure, it's the point of life.

I was open to whatever happened, not attached to any particular outcome. Of course, I hoped to get to Hawai'i, but by now I knew enough about oceans that I understood my success depended on a huge number of variables, many of them outside my control. So I decided to take a fatalistic approach – to simply keep showing up, one day at a time, get my backside on the rowing seat, and start sticking my oars in the water. Only time would tell if this would be enough to get the job done.

So courage, for me, was the product of faith, curiosity and openness – and maybe just a tiny dash of wanting to heal my injured pride. I used to think I needed courage *before* I could do something brave. Now I've flipped that belief. I believe we only find courage when we feel the fear and do the brave thing anyway, usually inspired by a cause or a community we care deeply about. We may feel great fear, but it's outweighed by our motivation, which is even greater.

Leaping Out of the Water

In his commencement speech at Kenyon College, Ohio, in 2005, David Foster Wallace told this story:

> 'There are these two young fish swimming along and they happen to meet an older fish swimming the other way, who nods at them and says, "Morning, boys. How's the water?" And the two young fish swim on for

a bit, and then eventually one of them looks over at the other and goes, "What the hell is water?"'[6]

Transcending paradigms is easy to say, not so easy to do. The first challenge is to get far enough outside the paradigm to see that we're in one. Our most deep-seated stories are often invisible to us. They can seem as much a fact of life as gravity, and as unquestionable. It might seem impossible, as if we have to transcend the paradigm before we can transcend the paradigm. Timothy Morton writes about *hyperobjects* – things that are so massive across time and space that we can't comprehend them. Paradigms are a kind of metaphysical hyperobject.[7]

The point is that we're usually so immersed in the norms and narratives of our culture that we don't even think about them. Some of them are pretty strange. Why does a Western professional woman, before she goes to work, put on expensive, uncomfortable shoes that will probably give her bunions and throw her back out of alignment? Why does her male colleague tie a brightly coloured noose around his neck? Why do we show our approval of a performance by slapping our hands together noisily? Why do we show affection by pressing our lips to somebody else's? Why do we believe that defecating and sex are bodily functions we should generally do in private, while eating or drinking in public is totally fine?

And then there are those things that only work because everybody has decided that they do, like money, religion, law and government. These are the stories that we have bought into so completely that we forget they're just stories. As the author of *Sapiens*, Yuval Noah Harari, wrote in an article for *The Observer* in 2018:

> 'Humans have always lived in the age of post-truth. Homo sapiens is a post-truth species, whose power depends on creating and believing fictions. Ever since the Stone Age, self-reinforcing myths have served to unite human collectives. Indeed, Homo sapiens conquered this planet thanks above all to the unique human ability to create and spread fictions. We are the only mammals that can cooperate with numerous strangers because only we can invent fictional stories, spread them around, and convince millions of others to believe in them. As long as everybody believes in the same fictions, we all obey the same laws, and can thereby cooperate effectively.'[8]

When we all believe the same story, it can be hard to remember that it *is* a story, especially when most people who believe the story really, *really* don't like it if somebody starts to question it.

Take Galileo, for example. His career had been going along swimmingly until he had the audacity to agree with Copernicus that the Earth moves around the Sun, rather than vice versa. The Roman Inquisition concluded that heliocentrism was foolish, absurd and heretical, forced Galileo to recant, and he spent the rest of his life under house arrest. If you expect to be thanked for speaking truth, you're likely to be disappointed.

The persecution of heretics is by no means relegated to the pages of history books. Our culture is complicit in suppressing perspectives, techniques and practices that challenge the mainstream narrative. Wikipedia labels therapies such as breathwork as 'New Age'. Of acupuncture, which has been helping the afflicted for 3,000 years, Wikipedia says: 'Acupuncture is a pseudoscience; the theories and practices of Traditional Chinese Medicine are not based on scientific knowledge, and it has been characterized as quackery.'[9]

Rupert Sheldrake's talk at TEDxWhitechapel about morphic resonance, the theory that creatures within a species share information in ways that defy conventional science, was removed from the TEDx YouTube channel for contravening TED guidelines.[10] Graham Hancock's talk at the same event, suggesting that the Amazonian hallucinogen ayahuasca can help humans access non-conventional sources of knowledge, was also removed. The journalist Lynne McTaggart's work on the power of intention, showing that groups of people meditating on peace creates a measurable short-term reduction in the amount of violent crime, has been dismissed as pseudoscience. Other public figures who dare to question the dominant narrative suffer character assassinations as conspiracy theorists, charlatans, spreaders of misinformation or just plain bonkers.

Of course, some of them possibly *are* nuts, but it is dangerous, dogmatic and narrow-minded to dismiss all narratives that run counter to the prevailing paradigm. As the joke goes: what is the difference between a conspiracy theory and a conspiracy? About six months.

My default position on such matters is that there is a spectrum, from completely true to utterly wrong, and most things lie somewhere in between. The truth is almost always more complicated than we would like it to be.

What We Don't Know

Former US Secretary of Defence Donald Rumsfeld famously said in 2002:

> 'There are known knowns; there are things we know we know. We also know there are known unknowns; that is to say we know there are some things we do not know. But there are also unknown unknowns – the ones we don't know we don't know … it is the latter category that tends to be the difficult ones.'[11]

As per the availability heuristic, we tend to give a lot more attention to the things we know than the things we don't. There is a much-quoted but probably mythical statistic that the average human has around 60,000 thoughts a day, of which 95 per cent are the same thoughts we had the previous day.[12] More credible and recent research at Queen's University in Canada puts the figure at 6,000 thoughts a day,[13] with no comment on the sameness or otherwise of those thoughts. Whatever the actual figures (and my guess is that there is a wide distribution curve for both thinkiness and repetitiveness), we can probably agree that, given that most of us have fairly repetitive lives, we are likely to have fairly repetitive thoughts. So it's not unreasonable to extrapolate that there might be vast expanses of reality where our thoughts never venture. I confess that I have never thought about what summer is like on Saturn, or what the Pope wears under his robes, or how jelly beans are made – until just now.

The following *What We Don't Know* model was inspired by the work of Louise LeBrun,[14] but I've made some additions.

First, let's admit to ourselves that there's a lot to be known in the universe. It is 13.8 billion years old, so although things were a bit slow for a while after the Big Bang (about a billion years before the first planets formed,[15] and 9.3 billion years before Earth appeared), that's still a lot of history. It is about 93 billion light years across,[16] and a light year is about 5.88 trillion miles. I'll let you do the arithmetic, but in short, it's a very, very big place. As to the number of celestial bodies, estimates vary, but let's assume a few zeroes don't matter between friends and go with figures of around 10 quadrillion stars[17] and 22 sextillion planets.[18] Chances are, some of them have something going on. So there's a staggering amount to be known about the universe, of which we know a vanishingly small proportion. Let's picture 'all there is to know' as a sphere, a very big sphere.

The further we push the boundaries of scientific knowledge, the more we know we don't know. Within the last twenty-five years, for example, we've discovered dark energy and dark matter. Dark energy constitutes about 68 per cent of the universe, dark matter 27 per cent[19] – that's 95 per cent of the universe that we've only just noticed, and know virtually nothing about.

Just imagine all we might know 100, or 200, or 500 years into the future, when our current level of knowledge will seem as primitive as the 1500s do to us now (noting that I have just fallen, yet again, into the linear progression fallacy).

So we can probably agree that, even though it sometimes feels like we're drowning in an ocean of information, there's still a lot we don't know. What we do know is a tiny fraction, a little skull cap sitting atop our imaginary sphere of all there is to know. Even when we include what we *know* we don't know – for example, I know that Mandarin exists, but I also know that I don't know how to speak it, any more than I can explain quantum physics, or build a rocket – I can be fairly certain the vast bulk of the sphere is still taken up by what I don't know I don't know.

Ballooning up above our skull cap, there is a whole other sphere of *what we think we know*; the gallimaufry of belief, religion, opinion, dogma, superstition and pet theories, and even our own memories that turn out to be faulty. We all, to a greater or lesser degree, mistake the contents of *what we think we know* for *what we know we know*.

The real test comes when we encounter something new and surprising, and what we choose to do with it. From our comfortable self-referential bubble of belief, buoyed up by confirmation bias, we often resist new ideas that are an uneasy fit with what we have already decided is true.

In *Liminal Thinking*, Dave Gray describes two tests that an idea has to pass before we incorporate it into our worldview:

> 'First, is it internally coherent? Does it make sense, given what I already know, and can it be integrated with all of my other beliefs? In other words, does it make sense from within my bubble? Second, is it externally valid? Can I test it? If I try it, does it work?'[20]

Unfortunately, we rarely get to Step 2 if we don't say yes to Step 1. The brain prefers coherence over accuracy, so if the new idea isn't consistent with our existing mental models, it will usually get rejected outright, as any climate change campaigner who has argued with a climate change denier – or vice versa – will know.

The more fragile someone's confidence in their existing worldview, the more staunchly they defend it. An attack on their beliefs becomes an attack on their identity, and open-mindedness shuts down.

So, all too often, we stick to our old version of reality, which has, after all, served us perfectly well all these years. We hastily construct a reassuringly solid barrier between our tried and trusted beliefs and the new and surprising information (sound of heavy door slamming shut in Truth's face).

Humanity is far from having unlocked all the mysteries of the universe, or even the mystery of our own brains, so it would behove us to maintain an attitude of humility, curiosity, critical thinking and willingness to consider new ideas, even from unexpected directions and unlikely messengers. If failing paradigms are to shift, we need to admit that we don't know everything there is to be known. As Yuval Noah Harari points out in *Sapiens*, maps in antiquity used to pretend to be a complete representation of the world. Then, during the fifteenth and sixteenth centuries, Europeans began to draw maps that included empty spaces, blanks to be filled in. It was only when mapmakers acknowledged the existence of *terra incognita* that exploration really took off.

Likewise, when Dmitri Mendeleev dreamed up the periodic table of elements (literally *dreamed* – the table's structure came to him as he was sleeping),[21] he left spaces in the table for as-yet-unknown elements, whose properties he could predict from their position in the chart. Within a few years, germanium, gallium and scandium were discovered, filling in the blanks.

So the burning question is: what do we do with these surprises, these sudden glimpses of a previously unimagined *terra incognita*? Are we willing to pay attention to them, and see where they might lead? Or do we say, 'nah, just one of those things', and move on? Do we want to stay cocooned in the ignorance of our bubble of belief, or are we open-minded and curious enough to shine the spotlight of our attention on the anomaly?

Hint: just about every major scientific or societal breakthrough has come from paying attention to the anomalies. Philosopher Arne Klingenberg writes, 'Anomalies essentially prove that the impossible is indeed possible.'[22]

If our situation seems inescapable, if it appears that humans have painted ourselves into an existential corner from which there is no apparent way out, could an anomaly potentially offer us an escape hatch?

Believing the Impossible

In our materialistic, supposedly scientific culture, there is a tendency for anecdotes about inexplicable happenings to get lumped together in the bucket of spiritual-supernatural-psychic-conspiracy-doolally-woo-woo: crop circles, UFOs, spontaneous remissions of cancer, intuitions, telepathic communications, ghosts, déjà vu, near-death experiences, extrasensory perception, to name but a few. I could relate any number of second-hand stories here, but I would prefer to draw on my own experiences, because these are the only ones I know for sure happened. My stories may be unspectacular by comparison with some, but they were enough to challenge what I thought was possible, and we can't all get abducted by aliens (nor do I have any particular desire to do so – just for the record).

Apart from the everyday magic of serendipity and synchronicity, three specific incidents disrupted what I thought I knew.

Ghostly presences: Writing in my journal one day, I idly wondered what it might be like to encounter a ghost, then remembered I'm really rather scared of ghosts, so I thought better of it, and let the universe know I was cancelling the order. That night in bed I was suddenly wide awake, with the distinct impression that somebody had just whispered in my ear. In the morning I mentioned the incident to my partner. He told me that he had been disturbed repeatedly during the night by multiple presences in the bedroom. Nothing like that had ever happened to either of us, before or since.

Metaphysical kisses: During the early months of a romantic relationship, one day we were kissing on the sofa and I had the sensation of rising up out of myself, and my energy connecting with his energy several feet above our heads, a feeling so intense that I pulled back in surprise. He asked, 'Did you feel that too?' and described a sensation identical to what I had just experienced. I can best describe it as a moment of extreme metaphysical connection.

My father's visiting card: On my 53rd birthday, in 2020, my mother sent me a birthday gift of three blank notebooks to use for my journal. I was moving some books off a shelf to make space for them when I noticed a small rectangle of card on the floor. I was astonished to find that it was one of my father's visiting cards. As a Methodist minister who averaged around ten house calls a day, he used these cards to share his contact details or to let people know he'd popped by in their absence. My father had died over sixteen years previously, and the card had the address and phone number for the house where we lived from 1970 to 1974. I'd moved house countless times since then, and the books and the bookcase were relatively new. I have absolutely no way of explaining how the card got there, but the fact that it was my father's *visiting card* seems self-explanatory.

We often conflate *inexplicable* with *impossible*. The left hemisphere loves certainty, but it can't always have it. There comes a point when continuing to insist on the impossibility of something starts to sound like wilful blindness. The right hemisphere is more comfortable with ambiguity, and there are times when admitting we don't know is the only credible response. As Mark Twain quipped, 'It ain't what you don't know that gets you into trouble. It's what you know for sure that just ain't so.'

Faith, Curiosity, Openness

I'm not asking you to be gullible. I'm not asking you to be so open-minded that your brains fall out. I hope I maintain a healthy sense of scepticism and that my bullshit detector is well maintained and fully operational, and I trust that yours is the same.

It takes discernment to know what and who to believe. It's all too easy to believe people based on how good they are at projecting confidence in their proclamations. Confidence is infectious – if they appear sure, we tend to assume that they have good reason to be – but it may not be so. When we're feeling wobbly, it's tempting to look to somebody who appears to know what they're doing. We might defer to authority, seeking out a father (or sometimes mother) figure who we believe will keep us safe from harm. We might even believe that they can turn the clock back to a time when the world felt more solid and dependable, that they can make all the uncomfortable uncertainties go away, that they can make our country great again. This is understandable, but also dangerous. If a leader says they are confident about the future, they are lying to you, and possibly to themselves. The future has never been so unknowable as it is now, and if we think we know it, we're delusional.

I'm inviting you to relax into the not-knowing, to be open to new possibilities, to be humble enough to hold apparently contradictory perspectives without feeling the need to decide prematurely one way or the other. It's when we know we don't know what we're doing, when we're well and truly in *terra incognita*, that we open up a load of new potentialities that we won't see if we think we already know the answers.

If we *can't* get okay with uncertainty, my one certainty is that we will screw this up. If we *can* get okay with uncertainty, we may still screw it up, but at least it won't be because we were too dogmatic and blinkered to see what is, rather than what we wish to be.

Chapter 10

Chaos

> 'Amidst every breakdown was the chance of a breakthrough. Times of great change were times of great possibility. Zadie thrived in such times, when the whirling coin toss on the fate of humanity hung in midair, when the slightest breath of kindness could sway the results. Anyone could be the person to tip the scales of the world. The outcomes of this pivotal moment could be determined by any of us. Everything rested on the choices of more than seven billion people. Each person's life mattered to the way the coin fell. One single act could alter the course of human history ... indeed, could ensure that humans had a future at all.'[1]
>
> Rivera Sun

Ocean eddies can be real buggers. My worst run-in with eddies was when I was rowing the second stage of the Pacific, attempting to cross the Intertropical Convergence Zone (ITCZ), historically known as the doldrums. The ITCZ, a band of low pressure that encircles the Earth near the equator, causes the wind to drop away to a breath, and then to nothing, making the doldrums the bane of sailing ships of old. Vessels could be becalmed for weeks on end, the sailors sweltering in the heat while their supplies of food, fresh water and rum ran low.

The ITCZ overlaps with the Equatorial Counter Current. You probably know that water goes down the drain in a clockwise fashion in the northern hemisphere, anticlockwise in the southern. The oceans echo the pattern, on a vastly larger scale. The entire North Pacific is one big clockwise swirl – north past Japan, eastwards across the Arctic Circle, then south along the western coast of Canada and the US – while the South Pacific circles anticlockwise. In between the two oceanic whorls is the counter current, an eastward flowing band of water between two western flows. Where water flowing one way meets water flowing the other way, eddies form – smaller circular currents that can really spoil an ocean rower's day.

These eddies defy plotting or prediction. There are so many variables at play in the ocean – not just winds at the surface, but also salinity, water density, the pull of the moon, the Earth's rotation, islands and land masses and the topography of the ocean floor – that we are even less capable of modelling ocean currents than of modelling the weather. And, unlike winds that let us know by their touch on our skin how strong they are and where they are going, eddies lurk unseen and unseeable. A seafarer can't detect an eddy until they are in it and they notice their boat is no longer going where they want it to go. Eddies sent me spinning in circles, and nearly drove me insane.

The ocean is a consummate chaotic system – complex, multivariable and unpredictable to the point of capriciousness. But not even the ocean is as chaotic as human civilisation.

Wicked Problems

In a complex adaptive system like the current human world, everything is interconnected, often in ways that are far too intricate to comprehend. The ability of a seemingly insignificant circumstance to generate disproportionate effects is simply stated by the centuries-old proverb:

> 'For the want of a nail the shoe was lost,
> For the want of a shoe the horse was lost,
> For the want of a horse the rider was lost,
> For the want of a rider the battle was lost,
> For the want of a battle the kingdom was lost,
> And all for the want of a horseshoe-nail.'

You have probably heard of the butterfly effect, a phrase coined by meteorologist Edward Lorenz to express the notion that a small change in an initial condition can have disproportionate and entirely unpredictable consequences further down the line: 'If the flap of a butterfly's wings can be instrumental in generating a tornado, it can equally well be instrumental in preventing a tornado.'[2]

Complex adaptive systems are a subset of chaotic systems, in which the behaviour of the whole can't be deduced from the behaviour of the components, and the components have wills of their own – like humans. As the number of human beings has swelled, and technology has amplified the geographical reach and interplay of our thoughts, we have become enmeshed in an ever-denser web of connection.

Our problems are not local and separate, but interwoven with a whole load of other problems to create so-called *wicked* problems, which are defined by the philosophers Alan Watkins and Ken Wilber as follows:

'1. A wicked problem is multi-dimensional.
2. A wicked problem has multiple stakeholders.
3. A wicked problem has multiple causes.
4. A wicked problem has multiple symptoms.
5. A wicked problem has multiple solutions.
6. A wicked problem is constantly evolving.'[3]

In short, a wicked problem is a kind of monstrous multi-tentacled creature, like a brain tumour that has sent out protuberances into all the most important parts of the brain, and those brave enough to try and solve a wicked problem are like the poor neurosurgeon tasked with removing the tumour without causing irreparable damage. Oh, and the tumour is continuing to grow visibly even while the surgeon operates.

As complexity increases, more and more of the problems faced by a system become wicked. When we look at the list of features above, we see that they describe much of what is happening on the world stage: climate change, economic inequality, racism, war and all the multitude of maladies that ail us.

As an extra sting in the tail, wicked problems do not just evolve; they often evolve *in response to* the very actions intended to resolve them. The line between problems and solutions becomes blurred; apparent solutions in a complex system can create unforeseen problems, and apparent problems can unexpectedly provide solutions. Scott Page, social scientist and a professor at the University of Michigan, says:

> 'An actor in a complex system controls almost nothing, yet influences almost everything.'[4]

A recent United Nations report, called *Interconnected Disaster Risks*, identifies the relationships between an assortment of global crises that fed off each other in 2020–21, including Amazon wildfires, an Arctic heatwave, an explosion in Beirut, floods in Vietnam, the Covid-19 pandemic, a cyclone, bleaching of the Great Barrier Reef, the Texas coldwave, the extinction of the Chinese paddlefish and a plague of locusts. Yes – *a plague of locusts*. The report may sound like a cross between the Bible and the script for a particularly far-fetched doomsday movie, but one of the lead authors, Dr Jack O'Connor, says:

> 'What we can learn from this report is that disasters we see happening around the world are much more interconnected than we may realize, and they are also connected to individual behaviour. Our actions have consequences, for all of us. But the good news is that if the problems are connected, so are the solutions.'[5]

The report mentions in passing that media reporting rarely goes deep enough to draw the connections to underlying causes, harking back to the WWF/ZSL report:

> 'Root causes are the underlying factors that create conditions for disasters to occur. If we think of an event such as the Texas cold wave as an iceberg, the unusually freezing temperatures that led to power outages and suffering were just the tip of this iceberg. However, this tip is how we perceive disasters, and this is [what] the media and discussions usually tend to focus on. Far below the tip, there are deeper structures that allowed the disaster to occur, and they are surprisingly similar for many seemingly unrelated events.'[6]

The authors define these 'deeper structures' as human-induced greenhouse gas emissions, insufficient disaster risk management, and undervaluing environmental costs and benefits in decision making. I hope it's becoming clear by now, and as I'll explore further anon, I believe there are even deeper levels to the iceberg: the narratives and values that form the foundation of our current civilisation.

When I was learning about complex adaptive systems, it occurred to me that this is yet another good reason to overcome our swing towards a left-hemisphere-dominated world. The left hemisphere loves to take things apart and then put them back together again, thinking that it now understands the whole because it understands the parts. It has a narrow spotlight that it shines on the problem at hand. It's not paying attention to the impacts of its solution on the world around the problem, nor in the future. This reductionist approach works well with cars and washing machines. It doesn't work so well with human societies or natural ecosystems.

We need the talent of the right hemisphere for seeing the bigger picture. The left hemisphere quite literally can't see the wood for the trees; here in the UK, developers of a proposed high speed train link have promised to plant new trees to replace the ancient woodlands that will be cut down, ignoring the fact that a wood is a deeply integrated community evolved over centuries, not just a bunch of trees you can cut down in one place and replace with saplings somewhere else. The right hemisphere intuits the wood as a place of harmony and rejuvenation (think of Japanese forest bathing), peopled by ancient trees and rich in diversity and relationship. The left hemisphere just sees a collection of individual, substitutable objects.

The left hemisphere has no hope at all of wrapping its head (so to speak) around the increasingly complex adaptive systems that characterise our planet. And yet our world is becoming more and more left-brain dominant – a dangerous combination.

Complexity might sound intimidating, a place of weird magic and anarchy and chaos, where reassuring rules of linearity and predictability break down and we're caught up in a tornado of uncertainty. You could be forgiven for swearing and stomping off in a huff, declaring it is all too much, and you may as well give up trying to do anything helpful in the world if it's all so hopelessly hit and miss. Better to do nothing at all than do something and find it had unintended consequences.

Unfortunately, this isn't an option. Like it or not, we're all in the chaos. Even though it may not be visible, everything we do has impacts – as does everything we choose not to do. No action is still action. No choice is still a choice.

I understand this can be a deeply unsettling thought for some folks. We can deny it, and do our best to build a sense of security based on money and home and material possessions, and for a while our bubble of belief will keep us feeling safe. We might even get away with living out the rest of our lives this way.

Or we can embrace the chaos, and dance with it to generate positive change. This option sounds much more exciting to me. So I'm going to try and show how we can turn these chaotic phenomena to our advantage in creating a new paradigm.

The Edge of Chaos

The edge of chaos is the fine line between order and disorder, towards which all complex adaptive systems tend to evolve. It's the balancing point where the components of a system are fluid, in motion, but haven't dissolved into complete tumult. The phrase was first used by computer scientist Christopher Langton, inspired by a scuba dive at the edge of the continental shelf off the coast of Puerto Rico. He describes his realisation that his previous diving, which at the time had seemed quite adventurous, was in fact very mundane when compared with the richness of life right on the edge of the continental shelf, at the convergence of the shallows and the deep, where life abounded – maybe, even, originated.[7]

In terms of human systems, the edge of chaos is the boundary between the status quo and anarchy, the zone of creativity where the system conducts its evolutionary experiments. The edge-of-chaos philosophy is actively embraced by some companies, such as Netflix. Its CEO, Reed Hastings, said in a 2020 interview on the Freakonomics podcast:

> 'All of Netflix is managing on the "edge of chaos". You want to be right up to that edge where it's dynamic and there's freedom. It has not fallen into chaos, but it's kind of right on the edge of it.'[8]

He adds the caveat that this is not a suitable model for all organisations. We probably wouldn't want doctors and nurses to be operating on the edge of chaos (not intentionally, anyway), but for cutting-edge companies in innovative industries like entertainment, Hastings believes this is how they maintain their (so to speak) edge.

If Ophuls and Glubb are right, we might already be on the edge, even if we don't yet know it. We have a choice: we can scurry back to the failing systems that we know, or we can welcome the edginess and use it creatively.

We have a lot more room for manoeuvre when we play in the places betwixt and between. The edge might sound like an uncomfortable place to be, but as the saying goes, if you're not living on the edge, you're taking up too much room.

Tipping Points

We already know that humans are not good at imagining exponential growth; we seem to be better wired for understanding slow, incremental, linear change. To be fair, this is nature's default way of operating, but with notable exceptions, which can be dramatic, even traumatic. Maybe this is why we prefer the linear narrative – it's less terrifying.

Social theory has adopted a term from evolutionary biology – *punctuated equilibrium* – to describe long periods of stability interspersed with short, sharp periods of radical change. In some kinds of systems, a parameter that has been changing slowly for a period of time crosses a boundary and the change suddenly becomes a very apparent *phase transition*, like water turning to steam as temperature increases. If the parameter that crosses the boundary leads to self-amplifying change, the boundary is known as a *tipping point*, and the system will find it very difficult, if not impossible, to return to where it was before the boundary was crossed. The system is now stuck in its new normal. An article in *Nature*, called 'Abrupt Climate Change in an Oscillating World', states this rather dryly:

> 'When the system is driven over a bifurcation point where a current equilibrium ceases to exist, an abrupt and irreversible shift toward a different equilibrium can occur.'[9]

Or, as Malcolm Gladwell frames it more vividly in *The Tipping Point*:

> 'Look at the world around you. It may seem like an immovable, implacable place. It is not. With the slightest push – in just the right place – it can be tipped.'[10]

We hear regularly about tipping points in the context of environmental collapse, where the decreasing Arctic ice cap, deforestation of the Amazon, ocean acidification and the melting of the tundra create self-reinforcing feedback loops (also known as positive feedback loops, or the snowball or domino effect) that may cause long-lasting alterations in our ecosphere that may not make us extinct, but could certainly make life a lot less fun for several millennia. These feedback loops are a feature of complex systems, and by their nature they are virtually impossible to model and predict accurately, especially when multiple feedback loops interact with each other.

So that's the scary side of tipping points, but there is also a more positive aspect when we look at tipping points in terms of human society and the spread of new ideas.

You might be familiar with the diffusion of innovations curve, first popularised by Everett Rogers in 1962,[11] and referenced by Simon Sinek in his TED Talk, 'How Great Leaders Inspire Action'.[12] It's a typical bell curve, ranging from the innovators on the left extremity, through early adopters, early majority, late majority, late adopters and laggards. As Sinek says of the laggards, 'The only reason these people buy touch-tone phones is because you can't buy rotary phones anymore.'

On this curve there is a make-or-break point for any new technology. To the left of that point, it has failed to achieve critical mass. Not enough people are using it to make it useful to the customer or worth further development on the part of the manufacturer. It goes to the great graveyard of failed ideas, often through no fault of its own – it was just that the time was wrong.

But if it makes it across the chasm into mass adoption, then it's off to the races. The good news – if you're an entrepreneur or social activist – is that the tipping point happens way earlier than you might think, at around 16 per cent adoption (the total of innovators plus early adopters).

And there are ways we can tilt the odds in our favour. Community change specialist and author of *Changeology*, Les Robinson, outlines the following attributes an innovation needs in order to succeed:[13]

1. **Relative advantage**. It's obvious to anybody using the innovation that it's better than what they had before; when the iPhone came out, my Palm Pilot got kicked into touch pretty damn quick.
2. **Compatibility with existing values and practices**. We like change, but not too much all at once; ever heard of the Xerox Alto, 1973's prototype personal computer? No, neither had I.

3. **Simplicity and ease of use.** If it's not easy enough, forget it; it's actually amazing that the internal combustion engine took off, thereby keeping car mechanics in work for the foreseeable future.
4. **Trialability.** Try before you buy; not only does a test drive allow you to experience the car, but psychologists say it also creates an emotional bond between car and prospective purchaser, because we feel more trust towards things we have touched.
5. **Observable results.** If the user perceives significant benefit, they are more likely to recommend the innovation, and peers are more likely to be prompted to ask for information;[14] 'You're looking fabulous! Have you been on a diet?'

So, in short, if we're trying to spread positive new ideas or behaviours from our own small spot in the complex system of human society, it's not rocket science to find ways that make them an easier *hell, yeah!* for folks. And there's a chance, if we can cross the chasm and engineer a massive outbreak of common sense, that a new paradigm could become our new stable equilibrium, never to return to its previous state.

Attractors

I would suggest a sixth factor to add to Les Robinson's list: we're more likely to adopt a new technology, product or idea if it's being used by people we know, like and respect. We are creatures conditioned by our connections, so trusted messengers have a huge influence. These influencers often coalesce around a particularly attractive idea, organisation or location.

In 2001, Paul Ray and Sherry Ruth Anderson published the results of questionnaires suggesting that 50 million Americans, a little over a quarter of the adult population, were what they tagged *cultural creatives*:[15] people deeply involved in spirituality and social activism, identifying with values of equality, conservation of nature, altruism, optimism, curiosity and universalism. But many of them felt lonely and isolated, believing that such people were few and far between and not to be found in their social circle or in the media. That's certainly how I felt during my corporate years, when I didn't see anybody else in my peer group asking the same burning questions about what it meant to live a good life.

So we need a focal point to bring these individuals together. As implied by the name, an attractor is an element in a complex system that draws elements towards it. There are periodic attractors and strange attractors (I think I may have dated some of those), but I'm most interested in point attractors and their role in systems change. Glenda Eoyang, founding executive director of the Human Systems Dynamics Institute in the US, describes how this works in *Coping With Chaos*:

> 'Scientific data has shown that changing attractors is not an easy, intuitive task. It requires two complementary actions.
> - A seed for the new attractor must be created outside the existing attractor regime.
> - Random noise or shocks must be introduced into the existing attractor regime.'[16]

This is starting to sound interesting. Can we cultivate seeds for new attractors, so that when a shock happens, we have viable and exciting new ideas lying around and ready to go?

I picture the current system as a magnet with a strong electromagnetic field. While there are no other magnets around, the old paradigm reigns supreme. Any iron filings in its vicinity fall into an obedient pattern aligning with the dominant magnet's field. But if a new electromagnet hoves into view, a few of the furthest iron filings start drifting away to the newcomer. And if there is a sudden blip in the electricity supply to the original magnet, there is a mass defection of iron filings from the old to the new magnet in town.

In a complex system there can be an awkward transitional time of flip-flopping, technically called intermittency, as the two attractors fight for supremacy. It's as if both electromagnets have dodgy power connections, and the iron filings are drawn to one, and then the other, and sometimes pulled in both directions at once. This can be very confusing if you're an iron filing, but the instability is also a hopeful sign, a symptom of the old regime breaking down to make way for the new.

There are a few new magnets emerging, such as complementary currencies that operate alongside national currencies, with goals such as rewarding the regeneration of the Earth.[17] At the time of writing, these currencies are still on the margins, but they're attracting an interesting band of pioneers, mostly young, idealistic, egalitarian and spiritually inclined.

Regardless of the eventual success or otherwise of the currencies, they are serving to bring cultural creatives together to combine forces and are introducing the idea of an alternative system of incentives.

Also, we know that our mainstream economic model is susceptible to major shocks, and, according to the Belgian economist Bernard Lietaer, it will always be thus for as long as the economic monoculture predominates, because boom and bust is inherent in the way the model is designed. (Not coincidentally, booms and busts represent a golden opportunity for the wealthy to get wealthier, while the rest of us get poorer, which could explain why there has been little attempt to design them out of the system.[18]) According to the International Monetary Fund, between 1970 and 2010 there were 145 banking crises, 208 monetary crashes and 72 sovereign debt crises. So it's just a matter of time before there's another crash, no matter how smart our economists think they are. As Lietaer put it:

> 'Our money system is structurally brittle. It doesn't matter if you put a very clever guy or a stupid guy at the wheel. The clever guy will take half an hour to have an accident, and the stupid guy will take ten minutes.'[19]

So while I'm not suggesting that global financial crashes are a good thing – clearly, they cause untold suffering – systems, human or non-human, seldom change without a triggering event. Shocks to the system are what create evolutionary change.

In 2008, when the last crash happened, complementary currencies were mostly still small, local and relatively underdeveloped. When the next crash occurs, who knows what might happen? It could be just the window of opportunity for conscious currencies to go mainstream.

Self-Organisation

Reed Hastings likens Netflix's 'edge of chaos' ethos to a jazz band rather than an orchestra:

> 'If you think about a conductor and an orchestra, it's an incredible thing, the level of synchronization, the level of precision, and it creates great art ... And then, there's this renegade little branch of music. And people

are riffing off of each other and responding and everyone is highly skilled. To be able to do that, you need incredible practice and a great ear ... It's less on top-down precision. It's less on efficiency and order and synchronization. And so it's a little bit chaotic, also very beautiful in a different way.'[20]

Entities within a complex system interact with each other from the bottom up according to a relatively simple set of rules, rather than being orchestrated in a top-down, hierarchical structure. A murmuration of starlings wheeling at dusk, for example, might look complex, but actually only three simple rules are at play. If you want to geek out on this, there are videos online where coders demonstrate how these three factors interact to make simulated birds, known as *boids*, behave in a flock-like manner.[21]

1. Separation: each bird stays far enough apart to avoid colliding with its neighbours
2. Cohesion: each bird stays within a certain distance from the centre of the flock
3. Alignment: each bird lines up with birds nearby

We also see self-organising dynamics at work in evolution through natural selection. Nature is constantly experimenting by generating new genetic mutations, some of which will prove to be biologically favourable and do better in the reproductive stakes, others fare less well and die out.[22]

You may have heard of the Darwin Awards, dark-humoured awards given annually in recognition of those humans who have voluntarily removed themselves from the gene pool through an act of excessive stupidity (such as triggering a self-burying avalanche, miscalculating a cliff dive, going kite-surfing in a hurricane, or jet-skiing off the Niagara Falls), and hence contributed to the overall improvement (or at least avoided the degradation) of the gene pool. Needless to say, the awards are usually posthumous.

This is nature's way of constantly improving the gene pool, favouring the fittest, where 'fittest' doesn't necessarily mean the biggest or strongest, but the one that *fits best* within its ecosystem.

Thinkers such as Barbara Marx Hubbard have suggested that, as the only species that is aware of its own evolutionary orientation, humans are in a position to consciously guide that process:

> 'Our new scientifically-based evolutionary universe story has given us the insight that nature and we ourselves are evolving. There is a direction in this process toward more complex order, more awareness, and more freedom to destroy or to evolve. Many of us are working together toward something we have never seen on any scale before – a sustainable, evolvable, co-creative society in which each person is encouraged to do and be his or her best.'[23]

Like any other part of nature, humans have great capacity for self-organising. But as we saw earlier, when civilisations enter their final phases self-organisation gives way to hierarchies and a multiplicity of tiny rules that stifle our creativity. But at the same time, as we vacillate in anticipation of a bifurcation point, exciting new social technologies emerge. We're seeing this in democratic social technologies like sociocracy and holacracy, in proposals for more public engagement in the processes of governance,[24] in collective impact initiatives based on getting people out of their siloes and into broad cross-sector coordination,[25] and in the shift towards decentralised teal organisations like Buurtzorg Nederland, as described by Frederic Laloux in *Reinventing Organizations*.[26]

In *Team of Teams*, General Stanley McChrystal describes how the Joint Task Force in Afghanistan in 2003 was forced into a major rethink of its operations when it became apparent that the scrappy, self-organising, low-tech cells of Al Qaeda were running rings around the mighty behemoth of the American military. They realised that, if they were to compete with Al Qaeda, they would have to sacrifice *MECE* (mutually exclusive, collectively exhaustive) efficiency in favour of *messy* effectiveness, embracing the full diversity of their resources.

Stan McChrystal's description is reminiscent of the flocking birds – separation, yet also cohesion and alignment – and like nature, the task force had to give up on long-term planning and strategy, and instead rely on rapid prototyping, evaluating the success or failure of each experiment as they went:

> 'Little of our transformation was planned. Few of the plans that we did develop unfolded as envisioned. Instead, we evolved in rapid iterations, changing – assessing – changing again. Intuition and hard-won experience became the beacons, often dimly visible, that guided us through the fog and friction. Over time we realized that we were not in search of the perfect solution – none existed. The environment in which we found

ourselves, a convergence of twenty-first-century factors and more timeless human interactions, demanded a dynamic, constantly adapting approach. For a soldier trained at West Point as an engineer, the idea that a problem has different solutions on different days was fundamentally disturbing. Yet that was the case.'[27]

Feedback

This democratisation of society sounds promising, but whether it works or not will depend on if and how well we can organise our feedback loops. In jazz, the band members know immediately if a riff has worked – either it sounds great, or clashing and cacophonous. Life outside a jazz club is rather more complicated, and we need ways to find feedback that may not be as obvious as a discordant note.

Stan McChrystal describes how the effectiveness of the task force relied on reliable, timely data. Daily O&I (operations and intelligence) briefings lasted an hour and a half, and were mandatory for the 7,000 personnel across the region, enabling them to share crucial information and best practice. In the Joint Operations Centre (JOC) at their bunker at Balad, McChrystal's task force could see on an array of screens what was happening all across the region in real time. This enabled them to see which strategies were working, which weren't, and to adapt accordingly.

In *Finding our Way*, Margaret Wheatley explains the importance of feedback:

> 'All life thrives on feedback and dies without it. We have to know what is going on around us, how our actions impact others, how the environment is changing, how we're changing. If we don't have access to this kind of information, we can't adapt or grow. Without feedback, we shrivel into routines and develop hard shells that keep newness out.'[28]

Many of our current troubles arise due to deeply imperfect feedback on our actions. The consequences of our choices are largely invisible to us. For example, the systemic failures that led to the 2008 sub-prime mortgage fiasco and the resulting global financial crisis have not yet been resolved, at least in part because governments intercepted the feedback loops by bailing out the banks. The perpetrators of the crisis still took home outsized bonuses, so there was no meaningful penalty for the behaviour that caused massive hardship for so many people. In the absence of direct and appropriately painful feedback, there is every danger they will do the same thing again.

(Incidentally, according to a report by the International Monetary Fund,[29] banks that had lobbied the governments were more likely to take bigger financial risks, and also more likely to receive government bailouts when the risks went sour. Correlation, or causation? I leave it to you to decide.)

Another example is climate change, in which all of us in the developed world are complicit to some extent. But unless we go out of our way to watch depressing documentaries about ecosystem collapse, we don't see forests being cleared to make way for cattle or the methane emitted by food tossed into landfill. We don't see the melting Arctic or the sea level rise affecting small island nations. People in the developed world lack the feedback that might motivate us to make better choices, while the people in the less developed countries, who are confronted with the feedback every day, have little opportunity to communicate their plight to the ones causing it.

Imagine for a moment: what would it be like if we had a command centre for our world like the one the task force had? What difference would it make if we had trustworthy, unfiltered, real-time information about the consequences of our actions, individually and collectively, against key metrics?

At the moment, we buy things without knowing where the raw materials and labour have come from; we throw them away without knowing where they go; we communicate without knowing where our words might land; we travel to tourist destinations without seeing how our presence affects local economies and cultures. Everything we do makes a difference, but we don't know what that difference is. If I could use my smartphone to scan the barcode of a grocery item and immediately see a time-lapse video of all the resources that went into its production, packaging and transportation (and even its slaughter if it is fish, poultry or meat), might I make very different purchasing choices? (Although if I could see its supply chains, I may not have bought that smartphone in the first place.)[30]

Even better, what would it be like if we could get reliable feedback on our collective projects? As we prototype new ways of living, growing food, supplying consumer goods, travelling and so on, wouldn't it be powerful if we could get feedback on the net impact of the new versus the old? Maybe we could even re-engineer our financial system to provide this feedback, to reflect the true costs and benefits of our actions, in order to encourage prosocial and regenerative behaviours and discourage destructive activities.

This might sound like (organic) pie in the sky. We are drowning in information – apparently at the start of 2020, we had 44 zettabytes of data in the world[31] – but big data has so far been aligned with corporate outcomes rather than environmental ones. But we can be braver about getting feedback. Too often we fall into good old confirmation bias when it comes to metrics – we want the good news, not the bad. We tend to seek out the data that show us how we're succeeding in our goals, not failing in them. We look for evidence of the intended consequences, not the unintended ones.

Can we counter confirmation bias with actively seeking out *dis*confirmation? In her TED Talk, 'Dare to Disagree', Margaret Heffernan tells the story of Alice Stewart, a doctor in Oxford in the 1950s, who was trying to find out why there was a rise in the incidence of childhood cancers. Based on the information she received via detailed questionnaires, she saw a startling correlation between childhood cancer and mothers who had been x-rayed during pregnancy. See the full talk for the rest of the story,[32] but the point here is Alice Stewart's long-term collaboration with a statistician called George Kneale. He apparently said, 'My job is to prove Dr Stewart wrong.' He actively sought disconfirmation of her theories, and she thrived on his feedback.

According to Margaret Heffernan:

> 'It was only by not being able to prove that she was wrong, that George could give Alice the confidence she needed to know that she was right. It's a fantastic model of collaboration – thinking partners who aren't echo chambers. I wonder how many of us have, or dare to have, such collaborators. Alice and George were very good at conflict. They saw it as thinking.'[33]

Reed Hastings calls this 'farming for dissent',[34] and this brings us to heterogeneity.

Heterogeneity

Built in 1943 as a temporary wartime facility, MIT's Building 20 became known as the Magical Incubator because it spawned an exceptional number of ground-breaking developments. Hastily constructed, it became the overflow building for researchers from a wide variety of disciplines. The numbering of the rooms was eccentric, so people would frequently get lost and wander into offices to ask for directions, leading to random conversations and cross-pollination of ideas. The cheap construction – it was also known as the Plywood Palace – meant that the inhabitants would often take it upon themselves to modify the structure to suit their purposes, drill or hacksaw in hand. It was chaotic, anarchic, heterogeneous, self-organising and unbelievably creative. And now, as of 1998, demolished, a mere fifty-five years after its supposedly temporary construction.[35]

Diversity was essential to this vibrantly creative chaos, and the need for diversity carries over into all kinds of contexts. The literature on corporate leadership is replete with examples of companies like Kodak and Blockbuster that failed to see a juggernaut of change heading their way because all the decision-makers shared virtually the same educational background, world-view and sources of information. Diversity of thought might have helped them see what was coming. There is a danger of the same happening in politics as career politicians come to monopolise the corridors of power, unlike previous eras when many had first careers in business, the military, education, or had otherwise experienced the real world outside the political bubble.

Nature knows that diversity leads to resilience. The greater the number of species, the greater the likelihood that at least some of them will make it through a sudden change in the environment. Humans are gradually catching up, but it depends on individuals getting comfortable with difference, and developing the capacity to tolerate opposing views.

Roberto Assagioli was the founder of psychosynthesis, a form of psychology based on finding meaning and purpose by connecting the deeply and uniquely personal to the transcendent and transpersonal. He believed that a mature personality required the individual to be able to hold contrasting poles in dynamic relationship, dancing between and synthesising apparent opposites in response to the unfolding circumstances, and that a mature society should also reflect this ability to embrace contrast without feeling the need to opt for one extreme or the other.[36]

His view was that the evolution of a society's consciousness depended on the participation of many people from a variety of backgrounds, with each person being free to be themselves. There is a slowly growing awareness in the business world that diversity is invaluable in avoiding groupthink, and not just a matter of paying lip service to political correctness, although this awareness has yet to reach many corporate boards. An EY report in March 2022 showed some improvement in ethnic diversity, but still only six CEOs in the FTSE 100 came from a minority ethnic background.[37] Gender balance is also slow to make headway: in 2018, excuses given by representatives of FTSE 350 companies on the lack of female representation included, 'We have one woman already on the board, so we are done – it is someone else's turn.'[38]

In human organisations, as in nature, homogeneity is fatal to resilience, so a wise management team overcomes its natural preference for seeking sameness. Organisations thrive when they embrace a diverse and co-creative dynamic. Nature banks on diversity, and is generous in her abundance of forms. Organisations can mimic the way nature allows a multiplicity of solutions to cooperate and compete.

This is called *coopetition*, meaning collaboration between competitors in the hope of mutually beneficial results. In *Complexity*, M. Mitchell Waldrop advocates using this principle to resolve disputes about the best way to achieve the goal: cooperate to agree the metrics of success, and then prototype all the competing methods to achieve it. The method that performs best against the metrics 'wins'.

Emergence

Emergence is the phenomenon in which the sum is greater than (or different from) the sum of the parts. It's pretty amazing. Co-founder of the Consilience Project, Daniel Schmachtenberger, describes it as 'the closest thing to magic that is actually a scientifically admissible term'.[39]

Let's take water as an example. As we all know from basic chemistry, water is made up of one part oxygen and two parts hydrogen, which are both gases at room temperature, but they come together to create H_2O, a liquid. How does that happen? In a further step of emergence, six or more molecules of water create 'wetness', a property that a single water molecule doesn't have.[40]

Some researchers believe that consciousness is an emergent property of complex brains[41] (although as we'll see later, this is hotly debated), so if they are right, the sense that I have of my self-ness is an emergent property of my neurons and the connections between them.

My personal opinion is that good ideas (or even bad ideas, come to think of it) are emergent. You might well have had the experience of mulling over a question for several days, weeks or months, until suddenly you get an insight, an *a-ha!* moment, usually when you weren't even thinking about the question but were in the shower, or driving, or some other such unfocused activity. I get a delighted thrill of surprise when this happens, a sense of astonishment that I had grappled for so long with something that now seems so blindingly obvious (the joy of hindsight bias).

But the surprise also comes from the realisation that the insight is more than the sum of the parts. All the pieces of information I've been gathering don't fully add up to the idea. Something else has been added, from I know not where. This, I think, is why the creative process has so often been described with reference to the divine.

Elizabeth Gilbert, the author of *Big Magic*, has described her belief that ideas are disembodied, energetic life forms with their own independent existence and consciousness, that roam the Earth in search of a human partner to bring them into manifestation. The idea's job is to be actualised. The human's job is to be open to the inspiration, and willing to download it into the material plane.[42]

The crucial elements, in my experience, are humility and patience. Inspiration rarely arrives on cue, and impatience will only drive it further away. It has to be wooed by creating the fertile void of uncertainty into which the thunderbolt of creativity can strike. Chaos and creativity go together. Certainty and solidity are fatal. As Margaret Wheatley writes:

> 'Knowledge is born in chaotic processes that take time. The irony of this principle is that it demands two things we don't have – a tolerance for messy, nonlinear processes, and time. But creativity is only available when we become confused and overwhelmed, when we get so frustrated that we admit we don't know. And then, miraculous, a perfect insight appears, suddenly ... Great insights never appear at the end of a series of incremental steps. Nor can they be commanded to appear on schedule, no matter how desperately we need them. They present themselves only after a lot of work that culminates in so much frustration that we surrender. Only then are we humble enough and tired enough to open ourselves to entirely new solutions.'[43]

Emergence can operate in groups as well as in individuals. Otto Scharmer's Presencing Institute, based out of MIT, uses co-sensing in working groups to generate creative new ideas. Their Theory U methodology involves letting go of preconceptions (what we think we know) in order to open up to new insights.

This approach was in part inspired by a conversation that Scharmer's colleague and collaborator Joseph Jaworski had in London with the physicist David Bohm, concerning the implicate order of reality. Bohm took two cylindrical jars, one slightly smaller than the other, with the smaller one having a crank on top. He placed the smaller cylinder inside the larger one, and filled the space between with viscous glycerine. When he placed a single drop of ink in the glycerine and turned the crank, the ink was drawn out to a fine ribbon until it seemed to disappear. When he reversed the motion of the inner cylinder the ink returned to its original state as a visible drop.[44]

What David Bohm was illustrating here was that we think the object, or ink drop, ceases to exist when we can no longer see it, but it does exist – it has merely transitioned from the explicate to the implicate order, from the obvious to the subtle. As Joe Jaworski writes:

> 'All matter and the universe are continually in motion. At a level we cannot see, there is an unbroken wholeness, an "implicate order" out of which seemingly discrete events arise. All human beings are part of that unbroken whole, which is continually unfolding. Two of our responsibilities in life are to be open and to learn, thereby becoming more capable of sensing and actualizing emerging new realities.'[45]

Our left hemisphere might be freaking out at this point. Humility and leaps of logic aren't its bag. It enjoys control and linearity, but clearly those are incompatible with emergence. And if we are to find the miracle necessary to escape our predicament, we will have to overcome the left hemisphere's wish for predictability and take a leap into the unknown, trusting our right hemisphere to intuit the way forward.

To summarise, we need to be right on the *edge of chaos* if things are to change. If we're too far back from the edge of the cliff, we won't bother to make changes. Once we're in freefall over the cliff, we'll have other things on our minds.

We are already in a *phase transition* – we can tell from the way things are flickering between one paradigm and another – so there's everything to play for in trying to make the new paradigm a good one, as once we cross the *tipping point* we could be in it for a while.

To optimise our chances of making the new paradigm an improvement, we need as many and as diverse (*heterogeneous*) people as possible to bring their gifts to the table, and specifically *not* the people who have held all the power in the old paradigm. They simply haven't needed to think about how things can be better, and as the beneficiaries of the old, they have no interest in designing the new; it would be like asking a French aristocrat to design the guillotine.

To fully embrace the diverse talents of our global design team, we need to be *self-organising*, taking our inspiration from the expert in effective coevolution, Mother Nature.

And when we gather around the *point attractor* of our desire to create a better world, sharing and prototyping ideas and gathering *feedback* from our peers and from our experiments, we maximise the potential for the *emergence* of a new way of living that none of us could have thought of individually, but together our imagination becomes greater than the sum of the parts.

This might all sound like wishful thinking if you take the view that the world is material, linear and predictable in a Newtonian kind of a way. But what if it wasn't? We'll explore an alternative worldview in the next chapter. It blew my mind. If you're not already familiar with these ideas, I hope it will blow yours.

Chapter 11

Reality

> 'If the doors of perception were cleansed every thing would appear to man as it is, Infinite. For man has closed himself up, till he sees all things thro' narrow chinks of his cavern.'[1]
>
> William Blake

Like me, Donald Hoffman is a preacher's kid. And like me, he is fascinated by the nature of reality. Way ahead of me, he has a radical new theory of reality, based primarily on evolutionary biology and backed up with mathematics.

As a child, Donald was curious about the discrepancy between the version of reality he heard on Sundays from his fundamentalist father (creationism, omnipotent God, hellfire and brimstone) and what he was hearing on weekdays from his science teachers. Now a quantitative psychologist at the University of California, Irvine, he sets out his theory in his book, *The Case Against Reality*.

At least since Plato's allegory of the cave in around 380 BCE, humans have wondered about the true nature of reality, and suspected that what we perceive with our senses may not be the full story but is in fact a projection of a deeper reality. According to the allegory, there is a group of people who have spent their entire miserable lives chained up in a cave, facing a blank wall. They see shadows on the wall, cast by people passing in front of a fire. But the people and the fire are behind them, out of sight, so the shadows are all the reality they know. According to the story, there is a philosopher who isn't manacled like the others; he can see the full scene, and tries to tell the people that what they're seeing is just a shadow of reality, not reality itself. But they don't want to know. They're not capable of making the mental shift required to recognise that the objects they now see are more real than the shadows that they've been conditioned to recognise as reality.[2] As Mark Twain said, it's easier to fool people than to convince them that they've been fooled.

In other words, we're conditioned to believe that our reality is real, that we are a solid, self-contained human body moving through a world of physical objects that behave in predictable ways. Some of them move – people, cars, animals, clouds, waves – and some of them don't – furniture, books, mountains, trees, teenagers when it's time to get out of bed.

For a while now, the conventional explanation of consciousness is that it's an emergent property of the neurons in our brains. As we saw in the previous chapter, some things collectively have properties that they don't have individually, like water molecules and wetness. So, the theory goes, individual neurons aren't conscious, but collectively they are.

Unfortunately, nothing is ever that simple, not even water and wetness. As with all metaphysical questions, opinions vary, but we could argue that wetness is a perception, not an inherent quality. Water can do many things – flow, flood, freeze, thaw, boil, evaporate, swirl and so on – but it can only be wet when there are nerve endings leading to a brain to perceive its wetness.

It's like the old philosophical riddle: if a tree falls in the forest and there is nobody to hear it, does it make a sound? The answer is no, because the 'sound' of the tree crashing to the forest floor is just a shockwave travelling through air. It only becomes a sound when it hits an eardrum, a microphone, or something else that can reverberate in response to the shockwave, translating it into a perception of sound. Likewise, wetness isn't an intrinsic quality, but something you experience when you put your hand under a tap, jump into a pool, or sit on a wet bench and wish you hadn't.

Coming back to brain cells and consciousness, there have been numerous attempts to answer the aptly called *hard problem of consciousness*, of varying degrees of intellectual desperation. The essential question is: what gives you a sense of being you?

The Hard Problem

We tend to assume that consciousness, our sense of self, is in our head – or, more specifically, in our brain. We know humans don't do well if our heads get cut off (although we also don't do well without a heart, lungs, gut or most other body parts generally regarded as necessary), and because our eyes and ears are located on our heads, we tend to experience the world from a very head-centric perspective. It has been pointed out that if our eyeballs were in our kneecaps, we would probably have a very different perception of where our consciousness lies.

And yet, *where* in the brain is it – if, indeed, it is in the brain at all? Science has so far epically failed to locate our centre of consciousness. As neuroscientist V.S. Ramachandran (rather anthropocentrically) writes in *The Tell-Tale Brain*:

> 'How can a three-pound mass of jelly that you can hold in your palm imagine angels, contemplate the meaning of infinity, and even question its own place in the cosmos? ... With the arrival of humans, it has been said, the universe has suddenly become conscious of itself. This, truly, is the greatest mystery of all.'[3]

We know that we have a sense of self-ness, of being an organism experiencing sensory input from our nerve endings, eyes, ears, nose and tongue, with memories, opinions, preferences, loves, hopes and dreams. But science has searched high and low, and as yet has no convincing explanation as to how we're aware of being us – let alone *why* we have been endowed with such awareness, as a sense of self seems to serve no useful biological purpose.

Neuroscientists have done a good job of identifying neural *correlates* of consciousness; specifically, finding out which parts of the brain correspond to which thoughts and feelings, but remember the cognitive error – correlation does not establish causation, and activity in these neurons may not cause those thoughts and feelings any more than Nicolas Cage movies cause swimming pool deaths.

As Donald Hoffman explains, an alien observer might see people gathering on a train platform, (almost) invariably followed by the arrival of a train, and deduce that the gathering of people caused the train to arrive. What the alien would not know is that humans have useful things called timetables, which are the invisible missing link between the arrival of the passengers and the arrival of the train. Just because two things always happen in the same sequence does not mean that the first one causes the second one. There could be a third factor, like the train timetable, that connects the two of them.

Likewise, neuroscientists can put you in their fMRI (functional magnetic resonance imaging) machine and show you photographs, play music or prod parts of your body, and measure your corresponding brain activity, but they have yet to prove a causal link between what you're subjectively experiencing and what the fMRI shows is happening in your little grey cells. To date, there is no credible scientific theory that describes how brain activity creates your you-ness.

So Donald Hoffman wondered if we were coming at this from the wrong direction, and decided to flip the question on its head. If we can't show how matter gives rise to consciousness, could it be that consciousness gives rise to matter? His interim conclusion is that yes, it does, that reality is in fact a *network of conscious agents*, and that space-time is a product of consciousness, which evolution has created to help us navigate our way through life.

A Paradigm Shift from Matter to Consciousness

If you're not familiar with these ideas, that's quite a lot to take in. Donald Hoffman's theory does great violence to most people's intuitions about what is objectively real and what is not, but I invite you to recall What We Don't Know, and suspend disbelief for long enough to consider the theory on its own merits. Remember we used to think the Earth was flat. This theory is equally paradigm-shifting, and as with round-earthism, it could be that 500 years from now people will wonder what all the fuss was about.

Could this be what Einstein meant when he said, 'Reality is merely an illusion, albeit a very persistent one'?[4] Max Planck, the theoretical physicist, was blunt in his assessment:

> 'I regard consciousness as fundamental. I regard matter as derivative from consciousness. We cannot get behind consciousness. Everything that we talk about, everything that we regard as existing, postulates consciousness.'[5]

As was Erwin Schrödinger (he of the hypothetical cat):

> 'Consciousness cannot be accounted for in physical terms. For consciousness is absolutely fundamental. It cannot be accounted for in terms of anything else.'[6]

When three Nobel Prize-winning physicists converge on the same opinion, it's worth paying attention – but the truth is, we don't know the truth. Physicists used to think that Newtonian models explained everything, until quantum theory came along and messed it up. We now live in an odd crossover period, where quantum physics successfully describes the world of the very small, and general relativity successfully describes the world of the very massive. Both stand up impeccably in experiments, yet nobody has yet been able to reconcile the two theories. This matters in relation to places where they both apply, like the start of the universe and the centre of a black hole. So, essentially, some very fundamental questions about the nature of reality are still anybody's guess.

(Speaking of quantum theory, you might be wondering why I don't drag in observer effects and collapsing wave functions to support my case. Two reasons: first, as bongo-playing physicist Richard Feynman said, 'If you think you understand quantum mechanics, you don't understand quantum mechanics', and I'm not going to try to explain something when I'm under no illusion (or delusion) that I've grokked it. Second, seems like every metaphysical book I've ever read references quantum theory, so there are plenty of places you can learn about it from authors who either defy Professor Feynman's dictum, or at least believe they do. In this instance I'm going with Abe Lincoln's dictum, that it's 'Better to remain silent and be thought a fool, than to speak and remove all doubt.')

Living in a Simulation

We think we see, hear, touch, taste and smell reality, but actually, we don't. We're living in a simulation, but it is not a computer simulation (at least, I don't think it is, although some other people do).[7] It's a simulation that each of us has been carefully constructing since the moment we were born.

Think about it – our brain has no way of perceiving reality directly. There it sits, a blob of about 86 billion neurons and a similar number of non-neuronal cells, in the warm, wet darkness of our skull, comfortably bobbing in cerebrospinal fluid, essentially inside its own sensory deprivation tank.

The only way it knows anything is to receive information from our sensory peripherals – eyes, ears, nose, tongue, nerve endings – and do its best to create a map that is good enough for navigating the world. But just as a map is not the territory it represents – rather, it shows only the features we need to pay attention to in order to safely and successfully reach our destination – our mental map doesn't have to be an accurate or comprehensive representation of reality, just the useful bits.

If this idea seems a bit of a stretch, try these thought experiments:

Experiment 1 – Scale

Consider your vantage point in the world compared with that, say, of an eagle, or a woodlouse. A soaring eagle's eyesight, plus its high vantage point, allows it to see about eight times as far as a human. It can recognise a rabbit or other prey at a distance of up to 2 miles. Google fails to provide me with data on how far a woodlouse can see, but we can imagine that the world looks very different from its perspective than it does from ours, or the eagle's. Objects we step over with ease loom like cliffs for the poor little louse. Some humans might be scared of spiders, but the woodlouse has much greater reason for arachnophobia, and it's largely down to scale.

I once met a flat-earther, who was 'proving' the Earth is flat by taking photographs showing that the horizon lay parallel to a ruler. This was also a matter of scale. If his ruler had been 10 or 100 miles long he may have observed that it diverged from the surface of our planet.

To get subatomic for a moment: if I were standing in front of you right now, you would probably perceive me as a solid body (well, solid with a few squishy bits). But science tells us that each of my atoms is about 99 per cent empty space (ignoring for now things like electric fields, magnetic fields, cosmic rays, dark matter and dark energy that exist in the 'empty'). If I could look at the back of my hand and zoom down to the subatomic scale, I'd see entities like electrons and positrons, quarks and antiquarks, popping in and out of existence as fleeting fluctuations in the fabric of space-time. Happily for me, I'm blissfully unaware of this so-called quantum foam. It doesn't even tickle. But I'm certainly not as solid as you, or I, think I am.

So, in other words, the scale of our body – or the distance we can see from our usual vantage point – determines the scale of reality that is relevant and useful to us.

Experiment 2 – Senses

Some creatures have access to far more sensory information than we do. For example, birds can detect polarisation of light. Mantis shrimps have sixteen visual pigments compared with our three. Bats can echolocate. Polar bears can smell a seal from 20 miles away, or 3 feet under the ice.

Even a simple walk with your dog, as it sniffs around every tree, lamp post and other dogs' bottoms, will demonstrate that your pooch experiences the world in a very different way than you do. Dogs have up to 300 million olfactory receptors in their noses, compared to about 6 million in humans. Their noses are even constructed differently from ours – humans breathe in and out through the same passageways, so a smell has barely entered our nostrils when we blow it out again. Dogs have a fold of tissue just inside their nostrils that helps to separate these two functions, so when a dog exhales, the spent air goes out through a side door – through slits in the sides of their noses – which not only maintains uninterrupted smelling services, but actually creates a circulation of air that ushers new odours in.

That's just one small example of how poor and puny many of our human senses are, compared with our animal (and possibly vegetable) companions.

Experiment 3 – Scariness or Sexiness

From the moment we're born, our experiences start to shape our view of reality. For example, if we're unfortunate enough to be badly parented, we may be over-sensitive to facial expressions that could indicate anger because our personal history has conditioned us to err on the side of caution. We see anger where there is none, because a false positive is more useful to us than failing to notice the danger signs that a caregiver is about to fly off the handle.

Or if we suffer from ophidiophobia, an innocent stick lying on the path might for a moment trick us into thinking it's a snake. We're hyperalert to things that we have a strong aversion to – or indeed, a strong attraction to. Our mind can deceive us into seeing the thing we most fear, or the things we most desire.

Experiment 4 – Timespan

Do you ever look at an ancient grandmother tree and wonder what tales she could tell if she could talk? Or a mediaeval castle, or a mountain? Their timescales are as different from ours as the difference in physical scale between a human and that woodlouse. Or maybe you notice a tiny songbird's lightning-fast responses – when a bigger bird swoops in at the bird feeder, the wren has vanished almost before your eyes have registered the woodpecker. The wren appears to live in a speeded-up world that our senses can't keep pace with.

We know that trees grow, that mountains erode, that tectonic plates drift, but we can't perceive any of that happening at the timescale on which humans operate. Yet again, our limitations of scale, sense and speed let us down, and we realise our perceptions are entirely subjective. Our senses are useful, in that they kept the continuous reproductive chain of our ancestors alive, and will hopefully do the same for us, but they don't give us the full picture.

Experiment 5 – Adaptation

I can speak to this from personal experience. When I first stepped ashore after my first 100-day voyage on a tippy rowboat, the ground seemed to lurch under my feet. I'd got so accustomed to the pitching deck that my brain had adapted to the constant movement of the boat, so solid land now felt like it was moving.

In 1950, Professor Theodor Erismann at the University of Innsbruck performed an experiment on his hapless assistant, Ivo Kohler.[8] He made him wear a pair of specially designed goggles that turned his world upside down – literally. Mirrors in the goggles flipped the visuals reaching his eyes, so top became bottom, and bottom became top. You can watch the video of the result on YouTube.[9] The subtitles are in German, but are entirely unnecessary as you watch poor Kohler stumble around and generally make a fool of himself. Trying to defend himself while his boss pokes him with a stick, he defends high when the stick goes low, and vice versa. When Erismann tries to pour him a cup of tea, he turns the cup upside down and tea splashes everywhere. He bumps into just about everything. His attempts to put one foot in front of the other look like something from the Ministry of Silly Walks.

But after what must have been the longest ten days of his life, his brain had adjusted, and he could do most daily activities quite competently. (The pair did further experiments, including having someone ride a motorcycle through Innsbruck while wearing goggles that reversed right and left. This seems like a terribly bad idea.)

The point I'm eventually getting to here is that the brain is tremendously adaptable. In just ten days it can reverse our perception of what is up and what is down, if that shift enables us to defend ourselves from a mad, stick-jabbing professor, or to get tea in our cup instead of on the floor. What other adaptations might we have made, over the course of a lifetime, or even over the whole of human evolution, so that our senses serve us better?

Fitness Beats Truth

Donald Hoffman proposes that over the course of millions of years, evolutionary biology has favoured organisms that perceive reality in a useful way, rather than an accurate one. Computer simulations have proved repeatedly that, other things being equal, an organism maximises its evolutionary potential when it optimises for *fitness payoffs* such as fight, flight, feeding and f… (ahem) mating, compared with organisms that optimise for truth. In other words, creatures that see the world *usefully* maximise their chances of passing along their genes, while the creatures that see the world *accurately* are weeded out of the gene pool, joining the unhappy recipients of Darwin Awards. As Hoffman puts it, 'The truth won't make you free. It will make you extinct.'[10]

So, over the course of thousands of generations, human perception (and presumably that of other creatures) has incrementally crafted an interface between us and the true nature of reality. This interface is extremely effective, and enables us to manipulate our world in ways that are beneficial to our survival and procreation, but it is not reality. Remember that your brain is a deaf, blind thing sitting in your cranium, doing its level best to create a model of reality that will enable you to interact with your surroundings in a way that keeps you alive. It doesn't *know* reality – it's just guessing.

When asked if reality isn't actually real, would he be willing to throw himself in front of a fast-moving freight train, Donald Hoffman replies that he would not, because there's a good reason we evolved to perceive throwing ourselves in front of fast, heavy objects as a bad idea. He says, 'I take reality seriously, but not literally.'[11]

He uses the metaphor of a computer desktop. Maybe I have an icon in the top right corner of my desktop that is blue and rectangular and represents a folder. This is not to say there is an actual blue rectangular object that contains my documents and spreadsheets. The reality is that my data lies in the pattern of ones and zeroes on the hard drive of my computer, but if I had to manipulate the data at the level of electrical impulses (even if I knew how to), it would take me a very long time to get any work done. So successive generations of computer programmers have created this extremely useful interface that enables me to do my work efficiently – or at least, if I don't, it's not my computer's fault. The user interface on the screen conceals the reality of the electrical pulses that constitute my data.

The left hemisphere of our brains is perfectly happy with this state of affairs. It doesn't over-concern itself with truth. In fact, it rather prefers icons to the real thing. Iain McGilchrist writes:

> 'The left hemisphere is not in touch with reality but with its representation of reality, which turns out to be a remarkably self-enclosed, self-referring system of tokens ... But the most curious aspect of the story of the left hemisphere is yet to come. It turns on how a self-consistent system of signs, such as the left hemisphere's world, can come to seem more real than the lived world itself.'[12]

This might imply that our right hemisphere has a better chance of all-areas access to the real world, but I'll come back to that later.

Alternative Interfaces

I've already mentioned how other species experience reality through a very different interface than the conventional human one, but so do some humans.

Donald Hoffman offers the example of synaesthetes, the approximately 4 per cent[13] of humans who have some degree of crossover between their senses. Michael Watson is a chef who perceives flavours as physical objects that hover in front of him, invisible, but which he can feel with his hands as if they were real. The taste of mint translates into a smooth column of ice. Angostura bitters becomes a basket of ivy. Other synaesthetes experience numbers as colours, people as sounds, or different musical instruments as physical sensations in different parts of their body.[14] It's as if evolution is experimenting with novel ways to perceive reality to see if they enhance or diminish the usefulness of our user interface.

We might also want to give a shout-out to psychics, empaths, clairvoyants, spiritual healers and mediums here. While there have no doubt been charlatans over the centuries who have taken advantage of people's desperation to contact departed loved ones, to be healed, or to steal a march on the future, there's also a significant body of evidence of abilities that can't be explained through conventional understandings of the nature of reality, but could potentially be explained by some people having access to information that is inaccessible to most of us. It could even be that this information is all around us, but only a small percentage of people are able to tune into the right wavelength to pick up the signal.

In *An End to Upside Down Thinking*, Mark Gober has compiled a comprehensive directory of examples that confound our current scientific belief system – we could even call it scientific dogma, based on the enormous reluctance of most scientists to entertain anomalies. His list includes the Stargate Project,[15] a secret US Army unit that ran from 1978 to 1995, using army staff trained in remote viewing (the capacity to sense information from afar, using the mind) to locate military bases, missing personnel, crashed planes and spies. The upside-down thinking of the title is the idea that material reality creates consciousness. Gober's contention is that the only way to explain these apparently impossible phenomena is if we turn this model on its head, as Donald Hoffman does, and investigate the possibility that consciousness is primary and creates our perception of a seemingly material reality.

Idealism and Idealism

Idealism is an unhelpfully ambiguous word, due to a quirk of the English language. It can mean the yearning for an ideal world, often used in a disparaging sense to imply naïvety, as in *young and idealistic*. But it's also the name for a metaphysical philosophical perspective, in which reality is a mental construct arising out of consciousness (usually contrasted with materialism, the belief that matter is fundamental and gives rise to consciousness).

Idealism would, ideally, be called *idea-ism*, but the English language doesn't like the glottal stop this requires, so it stuck an *l* in the middle. Donald Hoffman calls his school of thought conscious realism, or a non-physicalist monism. But his philosophy is within spitting distance of idealism, which is how Brazilian-Dutch philosopher, Bernardo Kastrup, refers to his own worldview.

A former computer scientist who started his career at the CERN large hadron collider in Switzerland, Bernardo Kastrup has published many books on the subject of consciousness as the underlying basis of the universe. His philosophy is summed up in the cheekily immodest title of one of his books: *Why Materialism is Baloney: How True Skeptics Know There Is No Death and Fathom Answers to Life, the Universe, and Everything*. In one interview he says, 'I believe that the philosophy of materialism is an in-your-face, incoherent theory of reality. And I'm completely ready to repeat that to anybody.'[16] And he does.

If you're under the impression that consciousness-as-fundamental is a fluffy, New Age kind of perspective, I urge you to check out Bernardo's work. He is about as un-fluffy as it's possible to be. With his close-cropped salt-and-pepper hair and goatee, and a voice that can squeak from tenor into alto when he gets especially passionate about his subject matter, which is often, Bernardo is a frequent guest on podcasts and YouTube interviews. With ruthless rigour, he dissects materialism like a swashbuckling surgeon with an intellectual scalpel, and shows it to be severely lacking as an explanatory narrative.

He goes beyond the perspective I mentioned earlier, that our brain sits in the darkness of our skull, earnestly doing the best it can to create a useful model of an external reality. He goes so far as to say that everything, *including our brains*, is created by consciousness. He leans heavily on the philosophical principle of parsimony, or Occam's razor – or, in Einstein's phrase, make everything as simple as possible, but no simpler. Materialism requires both mind and reality, while idealism requires only mind. He writes:

> 'Realism: Reality exists outside and independent of mind; Idealism: Reality consists exclusively of mind and its contents. Notice that materialism entails realism but goes beyond it: it postulates not only that matter exists outside mind, but that mind itself is generated by matter.'[17]

He shares some interesting thoughts on why we perceive ourselves as separate beings even though, according to his worldview, we all exist in, and are connected by, consciousness. Think of consciousness like a river that flows from a source. As it passes over the riverbed it forms little eddies, or whirlpools, that become a distinct form and create the illusion of being something discrete from the rest of the river. The self is like the vortex in the centre of the whirlpool, turned in on itself, reflecting itself, imagining itself to be something separate from the rest of the stream. But you can't take a whirlpool out of the river.

Continuing the metaphor, he sees our sensory perceptions as being the rim of the whirlpool where it interfaces with the rest of the river. The rim is where the *self* meets what it perceives as *other*. Water close by, just outside the inward-facing surfaces of the whirlpool, is our subconscious – the information that rarely makes it into the conscious awareness of the self-reflecting vortex, but nonetheless affects the shape and behaviour of the whirlpool.

So, you may ask, how is it that we seem to have a shared experience of an external, material reality? Bernardo says that all our whirlpools are in the same river of information-consciousness, so although each of us has a localised experience in space-time, as a subset of the entire river, we are mostly subjected to the same shared flow of information.

To explain how we've got so confused about the relationship between mind and matter, he uses another metaphor – that of an artist painting a self-portrait, and then confusing the portrait for himself. It is as if we have created (painted) a mental model (portrait) of our world, which we can measure, analyse, predict and control. This is analogous to Donald Hoffman's interface concept. The portrait is just a painting – it's not the person that it represents, but we mistake the self-portrait for ourselves. In Bernardo's metaphor, the painter looks at his self-portrait, and says, 'I am that!' and then tries to explain himself by reference to the brush strokes and pigments on the canvas, by reference to the representation instead of the reality. We have, in effect, created the icon of the brain to represent our consciousness, and then we attempt to measure and analyse the brain-icon in order to explain why we are conscious. We have got it the wrong way around. Consciousness created the brain-icon. The brain-icon does not create consciousness.

So then why would brain damage affect our perception of the world, if the brain is just an icon?

> 'If the brain is a whirlpool of mind, the patterns of whose flow determine the qualities of subjective experience, then physical interference with the brain should indeed alter subjective experience insofar as it changes such patterns; that is, insofar as it messes with the whirlpool.'[18]

This is necessarily an over-simplified presentation of a few key points from Bernardo's work, and even if it's all rather befuddling at the moment, I hope it's enough to make you question whether materialism might indeed be baloney, and consciousness could be primary.

This is shaping up to be a fiercely-contested debate between materialists and idealists, and it's not just an abstract debate between highbrow philosophers with nothing better to do with their time. It has major implications for how we understand reality, and hence our place within it. Our metaphysical story matters, quite literally.

Whether or not we are conscious of it, we all tell ourselves stories about who we are, how the world works, and what life is for. This determines where we put our attention. Our attention determines how we choose to act. How we act determines the world we create. Our stories translate into our material reality. They *matter*, because they *become* matter. Our stories define what we believe to be possible, or impossible.

At least for now, these metaphysical debates can't be settled one way or the other, especially given that the current scientific model is birthed out of the materialist worldview and hence biased in its favour. I urge you to resist the left-brained urge for certainty and to tolerate ambiguity. Pending proof, though, I tend towards idealism, because it permits a narrative that is more exciting and empowering than plodding materialism – of which more later.

Russian Dolls of Consciousness

Not only is Donald Hoffman proposing that so-called reality may be no more than a handy user interface, he also suggests that consciousness may exist at multiple scales, meaning that I am conscious, but so are the subsets of me (cells, for example), and so are the supersets that include me (the hyperobjects of which I am a part, like humanity, or even the whole planet).

Donald Hoffman and his colleagues have created a mathematical representation of consciousness.[19] When they use this to mathematically represent the interaction between two consciousnesses, the resulting representation is also a definition of consciousness, and so on, in ever increasing levels of complexity. Consciousness therefore appears to be fractal, with the organism at each level having a degree of consciousness, but not able to access the consciousness of the organisms at the levels above or below, resembling a set of nested Russian dolls, each doll unaware of the consciousness of dolls larger or smaller than itself.

This theory is supported by the lived experience of split-brain patients, the poor unfortunates who not only had their corpus callosum surgically severed to treat the most severe forms of epilepsy, but were then subjected to a lifetime of experiments by inquisitive neuroscientists like Roger Sperry and Michael Gazzaniga.[20] There are cases of split-brain patients having two hemispheres with quite different personalities, for example, the woman whose right hand would fight with her left hand in the supermarket over what items to put in her trolley[21] (I have days like that too), or the student whose right hemisphere wanted to be a racing car driver, while the left wanted to be a draftsman.[22]

Yet, despite these apparently warring hemispheres, when these patients were asked if they felt any different after the surgery than they had before, they replied that, other than being freed from their debilitating seizures, they felt exactly the same. At first this answer came from the left hemisphere, which is the only hemisphere that can speak, so the researchers asked the non-verbal right hemisphere the same question, providing it with a pencil or a set of Scrabble letters so it could communicate. It produced the same answer: no difference. So even though there was now *no physical connection* between the two hemispheres, the subjective perception of the patient was that they were still a single consciousness.

So if consciousness is fractal, and perceives itself as unified even in the absence of physical connection, this could disrupt our perception of ourselves as isolated individuals. At the cellular level, what looks like a human body is comprised of 57 per cent non-human DNA, so numerous are the bacteria, viruses and fungi that inhabit our bodies. They're not just impartial passengers – they influence our behaviour, for example, by triggering the release of happy hormones when we consume their favourite food. We might think it's us craving a cookie, but chances are it's our gut bacteria being in the mood for something sweet. As Ed Yong writes in *I Contain Multitudes*:

> 'Every one of us is a zoo in our own right – a colony enclosed within a single body. A multi-species collective. An entire world.'[23]

It could be that all the bacteria, viruses, *et al.* in my microbiome have a consciousness, albeit a limited one, and that I am the sum total of, and yet hopefully more than, these consciousnesses. As in emergence, it could be that my whole is greater than the sum of my parts.

Could this same principle operate at the species-wide level? Rupert Sheldrake's morphic resonance theory states that natural systems 'inherit a collective memory from all previous things of their kind',[24] and that morphic resonance explains phenomena that we might otherwise call telepathy. He believes that a species shares a collective memory, which individuals from that species both draw from and contribute to. Sometimes called the hundredth monkey phenomenon, and related to Carl Jung's concept of the collective unconscious, this means that once a new discovery is made by a species in one part of the world, other members of the species may spontaneously make the same discovery, even if there is no apparent way that the information could have been communicated.[25]

His views have aroused much controversy – I mentioned earlier that his talk was barred from the TEDx YouTube channel. In *SuperSense: Why We Believe in the Unbelievable*, Bruce Hood brusquely dismisses Sheldrake's idea of morphic resonance for being based on 'lawless, weak evidence'. He unfavourably compares the morphic field with electric and magnetic fields, which are measurable and obey predictable laws, while morphic resonance cannot be measured and is more capricious in its occurrence.

But scientists are as prone to confirmation bias as the rest of us, and can be quick to turn upon anyone who is so audacious as to put forward a paradigm-shifting theory that would require them to rethink their own theories or put them at risk of losing their research grant. As we saw in the context of GDP, just because something can't be measured (yet) doesn't mean it's not real.

Moving up the scale, James Lovelock's Gaia hypothesis proposes that the Earth is a self-regulating, synergistic system whose living organisms interact with their ecosystem to optimise conditions for their continued survival. This sounds very much like collective consciousness at the planetary level. Could it be that there's something that it's *like* to be a planet – in other words, a planetary consciousness?

If that seems a radical idea, remember that we're not good at perceiving hyperobjects that lie outside our human scale. Could it be that we fail to perceive Gaia consciousness for the same reason that the woodlouse fails to perceive the entirety of my garden? The perceptions that have been most useful for our survival are those that relate to us as human-sized beings, with a human-length lifespan. We've mostly been able to assume that the Earth is a constant in our lives, so there's been little point in burning precious brain-calories to pay attention to what our planet is up to, any more than I expect my gut bacteria to have an opinion about how I choose to live my life.

Cosmic Consciousness

Donald Hoffman's theory then, taken to its logical extreme, implies that humans are but one layer of being within an overarching collective consciousness that extends all the way up to the entire cosmos. Here we're verging on the spiritual. In an interview with Adrian David Nelson for the Waking Cosmos podcast, Donald Hoffman doesn't shy away from this question:

> 'The theory says that yes, this goes on ad infinitum ... the mathematics allows that this could go on to infinity so there are conscious agents with an infinite range of experiences. So once we get to that area – conscious agents with an infinite variety of conscious experiences – now we are treading on the turf of spirituality. We're talking about infinite consciousnesses. But we have mathematical precision – we can actually start to prove theorems about these higher level consciousnesses and the relationship to finite consciousnesses.'[26]

If that feels like a step too far for your belief system, I'll come at it from a more pragmatic angle. Research by Kate Diebels and Mark Leary shows that an appreciation of the interconnectedness of all life, even if not going as far as infinite consciousness, is beneficial to compassion, altruism and psychological wellbeing:[27]

> 'A variety of philosophical, religious, spiritual, and scientific perspectives converge on the notion that everything that exists is part of some fundamental entity, substance, or process ... In two studies, believing in oneness was associated with having an identity that includes distal

> people and the natural world, feeling connected to humanity and nature, and having values that focus on other people's welfare ... The belief in oneness is a meaningful existential belief that has numerous implications for people's self-views, experiences, values, relationships, and behaviour.'[28]

I can't resist it – I'm going to have to quote Einstein yet again, because his words so perfectly describe this idea of connecting with the 'bigger us':[29]

> 'A human being is part of a whole, called by us the "Universe", a part limited in time and space. He experiences himself, his thoughts and feelings, as something separated from the rest – a kind of optical delusion of his consciousness. This delusion is a kind of prison for us, restricting us to our personal desires and to affection for a few persons nearest us. Our task must be to free ourselves from this prison by widening our circles of compassion to embrace all living creatures and the whole of nature in its beauty.'[30]

So while you may remain to be convinced that consciousness is nested in a hierarchy of ascending levels, rather than isolated and discrete as we in the West have been conditioned to believe, there are clearly upsides in choosing a narrative that embodies this new possibility, simply on the grounds that it makes us better, happier human beings.

The Existential Death Spiral

This new theory of being-ness can help us overcome our fear of death. And, given that death is inevitable (sorry, transhumanists), this would be a good thing. Daring to go where most scientists would fear to tread, Donald Hoffman proposes that, if his theory of consciousness is correct, and space-time is an illusion and consciousness gives rise to the phenomenon of matter, death may not be as final as Western secular culture would have us believe. He likens it to taking a break from a virtual reality game. When we take off the headset, we disappear from the game, but we still exist outside it. This echoes Joseph Campbell's metaphor:

> 'If the body is a light bulb, and it burns out, does that mean there's no more electricity? The source of energy remains. We can discard the body and go on. We are the source.'[31]

Similarly, the Dalai Lama describes death as simply a change of clothes.[32] If true, this would have major impacts on the story most non-religious people tell themselves about the finality of death, with all the anxiety that brings. We've bought into the story of our own smallness, our vulnerability, our mortality, our individualism, our aloneness in the universe. What if we believed there is no such thing as death – only transformation?

There's a growing volume of academic research that supports the ancient belief in reincarnation that originated in Buddhism, Hinduism and Sikhism. Ian Stevenson, the former Carlson professor of psychiatry at the University of Virginia, has documented hundreds of detailed case studies of children who were able to recall verifiable details of previous lifetimes, and often bore birthmarks or birth defects that mirrored the fatal wounds suffered by their previous incarnation. Collectively, his body of work presents a compelling case that we have an essential self, a soul, which cannot be destroyed but flows continuously, alternating between incarnate and dis-incarnate states.

Christopher Bache shares potent insights in his revelatory book, *LSD and the Mind of the Universe*. On weekdays, Bache was a professor of religious studies at Youngstown State University. On the weekends, unbeknownst to his colleagues and students, he took seventy-three high-dose LSD journeys over the course of twenty years, in an earnest quest for spiritual wisdom. On reincarnation, he writes that by repeatedly experiencing death and rebirth countless times during his LSD sessions, eventually the concept of death began to lose its meaning. Even though his physical body would be destroyed at the end of each incarnation, he came to know that it was impossible for the core of his being, his soul, to die. No matter how complete the destruction, the soul always rose again, renewed.

It doesn't matter whether or not you believe this to be true. It's easy to be sceptical. Personally, I struggle to remember things I've done in this lifetime, let alone a previous one. But I find it a reassuring belief. First, it alleviates my otherwise unscratchable itch to see how the future will play out, long after I'm dead – I like to believe that my future incarnation will get the chance for an update, although I almost certainly won't remember that in this present lifetime I was curious to know. Secondly, it's a powerful incentive to create a civilisation that works for everybody. Chances are that in previous incarnations you've been every colour, every gender, every nationality, every class, or if you haven't yet, in a future lifetime you will be. If this knowledge doesn't breed greater compassion, I don't know what will. It makes sense to desire a fairer world because we just don't know who we might be in our next incarnation. If we turn a blind eye to discrimination based on colour, gender, sexuality, religion, income bracket or any other equally arbitrary and oppressive distinction, we may well find ourselves on the receiving end of that very same discrimination on our next time around.

It also makes sense to want a more beautiful world, rich in biodiversity, clean air and pure water, because we're going to have to live in it in future incarnations. If we tolerate mass extinction, pollution, deforestation or desertification, it's going to be an impoverished world to which we return. This long-term perspective plays little part in our current civilisation. If we knew that this is our planetary bed and we're going to have to lie in it for a long time to come, we might feel a greater sense of responsibility towards future generations. We would try harder to be good ancestors.

Of more immediate consequence, compelling research suggests that fear of death makes us less happy and more materialistic.[33] Adopting a different belief around death could enable us to escape what I call the existential death spiral, a terror management strategy in which climate change (or any other existential threat) makes us anxious, so to divert ourselves and create a reassuring external representation of permanence and immortality we go shopping, which exacerbates the very problem that made us anxious in the first place, leading to still greater anxiety, more retail therapy, and so on.

As Bernardo Kastrup points out, a materialistic philosophy – in the sense that matter is primary – leads to materialistic behaviours, in the sense that we value material wealth. As with the double meaning of idealism – idealism as utopian wishfulness and idealism as a philosophy, we have a corresponding double meaning of materialism: if you believe that the world is made of matter and *only* matter, then acquiring more and more matter makes some kind of sense. There is a human impulse to want more of whatever you believe is important – more of what you believe matters, I suppose you could say. So if matter is all that matters, then more matter is what you want.

If death lost its sting, in other words, we might be less anxious, feel less compulsion to buy stuff we don't need, and improve the long-term prospects of our species. What's not to love about that?

Chapter 12

Ego

> 'We do not need a new religion or a new bible. We need a new experience – a new feeling of what it is to be "I".'[1]
>
> Alan Watts

In early 2020, near the start of the Covid-19 pandemic, I had the honour of interviewing Buddhist nun Jetsunma Tenzin Palmo via video call for my book, *The Gifts of Solitude*. I'd decided to write *Solitude* as my contribution to what felt like a pandemic war effort. I realised that many people were struggling with feelings of isolation during lockdown and, believing myself to be something of an expert on the matter, having spent 520 days and nights alone on the ocean, this seemed to be the most appropriate way for me to offer my gifts in service to the greater good. I was delighted, excited, and not a little starstruck, when Tenzin Palmo accepted my request for an interview.

However, the Monday morning of our conversation I felt dishevelled, my hair damp from the shower, my eyes strained after an intense weekend staring at the computer screen while I pounded at the keyboard. I was racing to write the book, fondly imagining that the lockdown might soon be over. I completed the first draft in seventeen days – but as we now know, I needn't have rushed.

The call connected quickly, with a solid internet connection to the nunnery Tenzin Palmo founded in northern India. She looked robust, healthy and serene – the exact opposite of how I felt. She was sitting in front of a white wall bearing a colourful cloth banner of a Tibetan deity. When she flashed her smile, which was infrequent but broad and dazzling, it reminded me of somebody. It was only later I was able to place it – Cameron Diaz, if Cameron Diaz were a shaven-headed, 76-year-old Buddhist nun from Bethnal Green.

It was early 2004 when I first read the book documenting Tenzin Palmo's life, *Cave in the Snow*. It had been recommended by a friend I'd met while travelling in Peru. He'd said there was an amazing book about a British nun who spent twelve years meditating alone in a cave in the Himalayas. I said it sounded incredibly boring.

But then I read it, and found an inspiring story of a brave, pioneering woman who had the audacity to aspire to attaining enlightenment while incarnated in a female body, a great rarity in the Buddhist tradition. This mission would involve not only unwavering dedication to her spiritual path, but doggedly ploughing her way through the institutionalised sexism and misogyny of organised religion.

I can't do justice to her story here, and highly recommend you read the book, but to give you an extremely potted biography: Diane Perry was born in east London in 1943, and realised she was a Buddhist at the age of 18 when she read *The Mind Unshaken*, by John Walters, which finally provided answers to her questions about herself and about life. Aged 21, she travelled to India and was ordained as a Tibetan Buddhist monastic, taking the name Tenzin Palmo. (*Jetsunma* was granted to her in 2008, and is a rarely bestowed honorific title meaning 'Venerable Master'.)

Having diligently pursued her calling in various Indian monasteries, in 1976 she felt called to find a more rarefied location for her meditation practice, and found a small cave – really more of a rocky overhang – in a remote location in the Himalayas, 13,200 feet above sea level, well above the snowline. Some companions helped her build out a wall to enlarge and protect the cave, but it was still tiny and cramped.

Life was harsh. She grew her own vegetables (she's a great advocate for the humble turnip), and in the summer supplies were delivered from a nearby village, although they didn't always arrive. In the winter, snow prevented access, so she relied on stockpiles. Temperatures often fell below minus 30 degrees centigrade. Once, her cave was completely buried in snow, and she had to dig her way out with a saucepan lid. She meditated for twelve hours a day in a meditation box. She slept in the box too, sitting upright.

The cave was her home for twelve years, from the age of 33 to 45. For the first nine years, she made occasional trips away or had visitors, but for the last three she was on strict retreat, with no human contact.

The book made a deep impact on me. Her story was one of the factors that inspired me to row across oceans. In my own, very different, way, I hoped that my solitude on the ocean might lead to, if not enlightenment, at least to some greater understanding of myself, life and the nature of reality. I thought of her often as I was rowing, especially to restore morale when I was feeling sorry for myself.

I thought I knew a thing or two about solitude. But I bowed down in awe of this woman.

Of all the wonderful wisdom Jetsunma Tenzin Palmo shared during that conversation, the gem that struck me most resoundingly came towards the end of our call, when she described the Buddhist Wheel of Life, or *samsara*, the endless cycle of life, death and reincarnation:

> 'At the hub of the wheel are three animals – a pig, a rooster and a snake. The snake stands for anger, the rooster for desire and greed, and the pig for ignorance. And they're biting each other's tails ... And their going round and round and round is what sets the whole wheel spinning. And that's the problem. Our delusion about our true nature, which gives rise to greed and desire for what gives pleasure, and aversion and anger towards that which does not give pleasure. And it keeps the whole wheel circling, no matter our best intentions. So the only way to stop the wheel is to break the hub.'[2]

I wondered out loud how we could stop this never-ending cycle of ignorance, greed and aversion. Tenzin Palmo's answer has stayed with me ever since:

> 'Within those three, the important one is our ignorance, our ignorance of our true nature. We identify with our ego, and although our ego is a good servant, it's a terrible master, because it's blind ... Once we realise that the nature of our existence is beyond thought and emotions,

> that it is incredibly vast and interconnected with all other beings, then the sense of isolation, separation, fear and hopes fall away. It's a tremendous relief!'

There's so much to explore in that one statement. Ego as trickster, blinding us to our true selves. Ego as a terrible master, echoing Einstein's dictum about the rational mind. Our true nature as vast and interconnected. Our false sense of isolation, giving rise to fear. It also reminded me of the left hemisphere and its grasping, manipulating nature, synonyms for greed and desire. It seems the Emissary and the pig have a lot in common.

Brain as Filter

What is the relationship between the ego, the brain and consciousness? Jill Bolte Taylor's TED Talk, 'My Stroke of Insight', offers some clues. It gives a mind-blowing account (so to speak) of her catastrophic left-hemisphere stroke at the age of 37, and is listed in the twenty-five most popular TED Talks of all time.

A professional neuroanatomist, she was able to observe her life-threatening predicament with (mostly) dispassionate calmness. As the blood wreaked havoc in her left hemisphere, shutting down its functions, she found herself seeing reality in an entirely different way which, far from being terrifying, was extremely alluring:

> 'As the haemorrhaging blood interrupted the normal functioning of my left mind, my perception was released from its attachment to categorisation and detail. As the dominating fibres of my left hemisphere shut down, they no longer inhibited my right hemisphere, and my perception was free to shift such that my consciousness could embody the tranquillity of my right mind.'[3]

She describes the sense of liberation that she experienced as being what Buddhists might call Nirvana, 'In the absence of my left hemisphere's analytical judgment, I was completely entranced by the feelings of tranquillity, safety, blessedness, euphoria and omniscience.'

She uses phrases that could equally be used by someone describing a peak meditative experience, or the effects of entheogenic drugs, such as *peaceful bliss of my divine right mind, glorious bliss* and *sweet tranquillity*.[4] The stroke disabled the filtering function of her left hemisphere, allowing her to pierce the veil of illusion.

> 'My left hemisphere had been trained to perceive myself as a solid, separate from others. Now, released from that restrictive circuitry, my right hemisphere relished in its attachment to the eternal flow. I was no longer isolated and alone. My soul was as big as the universe and frolicked with glee in a boundless sea.'[5]

This seems to validate Bernardo Kastrup's view that the mind generates a self-reflecting whirlpool that becomes blind to the universal (or at least, species-wide) stream of consciousness that lies beyond the boundaries of its vortex. The brain, as an icon representing mind, acts as a reducing valve on consciousness. Damage to the brain/icon, such as a stroke, acts like a branch thrust into the side of the whirlpool, disrupting its self-mirroring surfaces, allowing the mind to glimpse the infinite stream of consciousnesss beyond.

In *The Doors of Perception*, Aldous Huxley describes a similar disruption induced by mescaline, a hallucinogen, and concludes:

> 'To make biological survival possible, Mind at Large has to be funnelled through the reducing valve of the brain and nervous system. What comes out at the other end is a measly trickle of the kind of consciousness which will help us to stay alive on the surface of this particular planet.'[6]

This echoes Donald Hoffman's finding that fitness beats truth, and our consciousness is conditioned to pay attention to survival rather than reality.

It's a radical paradigm shift from what most of us were conditioned to think, and it certainly boggled my mind when I started to learn more about it, to understand *we are all one* as not just a spiritual truth, but a scientific one. For a moment, forget everything you think you know about how consciousness works. Just put your preconceptions to one side. You can pick them up again afterwards if you want to, if you don't find this alternative worldview convincing or agreeable.

Now, picture consciousness as being completely and utterly everywhere – filling the room you're in, permeating the chairs and tables, the floor, the ceiling, permeating you and every cell of your body. Picture consciousness filling your home, your garden, your village, town or city – all the buildings, all the people, everything. Picture consciousness sinking down into the Earth, all the way to its core, and all the way out into space, out through the atmosphere, across the void between planets, filling our solar system, the Milky Way, the whole vast cosmos. Picture consciousness enveloping all those quadrillions of stars in the universe, along with planets, moons, asteroids, comets, black holes, white dwarfs and red giants.

Now imagine what it would be like to hold *all that* in one human mind. It's inconceivable, right?

So could it be possible that we're part of that infinite cosmic consciousness, as a drop is part of the ocean, but in order for us to function in a day-to-day kind of a way, our brain has to filter out almost all of that vast consciousness apart from the infinitesimally tiny slice of it that relates to this body we call our own? Could it be that all of us are, as Jill Bolte Taylor describes, souls as big as the universe, frolicking with glee in a boundless sea?

This idea has a growing number of advocates. Bernardo Kastrup hypothesises that the function of the brain is to localise consciousness within time and space, confining the information available to what can be experienced from the perspective of an individual being. Without the brain performing this filtering function, awareness would extend across all time, all space and all dimensions. How mind-blowing would that be? Except that, presumably, infinite mind would have infinite capacity, and would remain nonchalantly un-blown.

In *How To Change Your Mind*, Michael Pollan cites the work of Dr Robin Carhart-Harris at the Centre for Psychedelic Research at Imperial College, London. There are various ways, including the use of psychedelics, but also breathwork, meditation, prayer and fasting, to deactivate the brain's default mode network (DMN). The DMN is the normal chatter that goes on in our minds when we're awake, daydreaming, remembering the past, planning for the future. Pause from reading now and stare into space for a moment. It won't be long before your DMN starts up.

When researchers scanned the brains of test subjects undergoing an intense psychedelic experience, you might expect to see brain activity going berserk, but instead, the DMN went quiet. According to Michael Pollan, this suggests that the DMN doesn't just control what information arises within, but also what is allowed into consciousness from the outside world. It operates as a filter or reducing valve to prevent the brain being overwhelmed by superfluous input, restricting incoming data to the minimum necessary to keep us alive.[7]

The interim conclusion seems to be that this filtering is in the interests of efficiency. We've already considered the brain as a prediction-making organ, and it can do this more efficiently if it takes in just enough information – but no more – to make an educated guess as to what will happen next. Michael Pollan describes the implications of this as 'deep and strange', our perceived reality being a seamless blend of external sensory input and our preconceived mental models, rather than the world as it is.[8] Stephen LaBerge, a psychotherapist specialising in the study of lucid dreaming, goes further, saying that while the dream state is dreaming unconstrained by sensory input, the conscious state is also dreaming, just constrained by sensory input. What we think is reality may be more accurately described as hallucination within certain sensory parameters.

Jill Bolte Taylor's experience illustrates why this hallucination may be more useful to us than unfiltered reality. As the stroke takes her filters offline, the beauty of what is revealed is so distracting that she keeps forgetting she's supposed to be calling for an ambulance. As she's dialling, her attention keeps being hijacked by the wonders of the universe, when she really needs to be focusing on the more urgent business of staying alive. So we can see why the brain has evolved the way it has – there are times when survival needs to take precedence over awe and wonder, when an ambulance is more important than amazement.

Eben Alexander, an American neurosurgeon who spent seven days in a deep coma after a rare form of bacterial meningitis attacked his brain's neocortex, wrote about his near-death experience in his book *Proof of Heaven*:

> 'We can only see what our brain's filter allows through. The brain—in particular its left-side linguistic/logical part, that which generates our sense of rationality and the feeling of being a sharply defined ego or self—is a barrier to our higher knowledge and experience.'[9]

We humans tend to be very proud of our brains, seeing them as what sets us apart from, even above, other animals. We've certainly paid a heavy price for our enlarged grey matter – human babies with their big heads have to be birthed long before they're ready to live independently, and until relatively recently childbirth was a major cause of death for both women and infants in developed countries – and remains so in many parts of the world. Naturally, we want to believe that these sacrifices have been worthwhile. We want to believe that brains enhance our human experience, rather than detracting from it. But it may not be so. Maybe our brains do indeed serve us well, by filtering out the aspects of reality that are not directly relevant to our survival. But in keeping us alive, they may also filter out much that is beautiful and wondrous and, in fact, extremely real, even if we can't see it.

But maybe we can connect with Mind at Large when we need access to more than the usual heavily-filtered content. In his book *The Surrender Experiment*, Michael Singer writes about taking an unorthodox approach to writing a crucial final paper. His obsession with meditation had interfered with his doctoral studies, and according to conventional wisdom he was woefully under-prepared. So he decided to use his meditative mind to tap into inspiration from the collective field. Without the help of books that would have brought him back into his intellectual mind, he started to write with what he calls 'the natural logic of a clear, unpressured mind'. Without self-censoring or editing, the words flowed forth, culminating with a flash of inspiration in which the purpose of the paper became crystal clear. He observed the entire process with detachment, marvelling at the powerful insights as they emerged onto the page as if guided by a greater intelligence than his own.

In other words, when he got the filters of his mind out of the way, he was able to access the field of knowledge and find the insights he needed to write the paper.

There are hundreds of examples that provide apparent evidence for brain-as-filter in books such as *One Mind: How Our Individual Mind Is Part of a Greater Consciousness and Why It Matters* by Larry Dossey, and *An End to Upside Down Thinking: Dispelling the Myth That the Brain Produces Consciousness, and the Implications for Everyday Life* by Mark Gober, as mentioned earlier. I won't repeat them here, but recommend those books if you want more case studies. *Infinite Awareness: The Awakening of a Scientific Mind*, by neuroscientist Marjorie Woollacott, also has excellent descriptions of near-death experiences that are inexplicable from a purely materialist standpoint.

If everything that is known exists in this shared field of consciousness, it might explain how clairvoyants can 'see' things in their mind's eye that they can't see with their physical eyes. It's one possible explanation for Rupert Sheldrake's experimental observations of dogs able to sense when their owners are coming home, even when the owner's departure time is randomly generated and they travel home by some means other than their own car. It could be the basis for *shared* near-death experiences, when those keeping vigil by a death bed have a collective experience of seeing a beautiful, bright light in the room as life leaves the body, or see other departed relatives waiting to greet the soul in transition.

I'm not saying that collective consciousness is necessarily the correct explanation for these phenomena, but it's one possibility. Conventional science can't even begin to explain them, so tends to sweep them under the carpet. After a while, all those pesky anomalies start making the carpet uncomfortably lumpy.

We can at least be humble enough to admit that we don't know. I'll leave you with this pointed question from Michael Singer:

> 'Am I better off making up an alternate reality in my mind and then fighting with reality to make it be my way, or am I better off letting go of what I want and serving the same forces of reality that managed to create the entire perfection of the universe around me?'[10]

Right, thought experiment over. The preconceptions you put to one side earlier – you can pick them up again if you like. Or maybe you prefer this view of the universe, and your place in it, in which case you can abandon those preconceptions. It's entirely your choice. Would you rather be the isolated, limited you, or the connected, cosmic you?

We Are All One

If our souls are as big as the universe, as Jill Bolte Taylor experienced, then could it follow that we are *all one soul*? Erwin Schrödinger certainly thought so:

> 'Inconceivable as it seems to ordinary reason, you – and all other conscious beings as such – are all in all. Hence this life of yours which you are living is not merely a piece of the entire existence, but is in a certain sense the whole.'[11]

Not only are we each a piece of the whole, but we are the whole experiencing itself. In *Conversations with God*, Neale Donald Walsch emphasises that *we are all one*, describing our sense of self as being a delusion that separates us from each other and from our spiritual source:

> 'Consciousness is a marvellous thing. It can be divided into a thousand pieces. A million. A million times a million. I [God] have divided Myself into an infinite number of "pieces" – so that each "piece" of Me could look back on Itself and behold the wonder of Who and What I Am.'[12]

In the 1800s, the German philosopher Arthur Schopenhauer wrote of 'that one eye of the world which looks out from all knowing creatures'.[13] He believed that every living being is the 'pure subject of knowing', universal and immortal, which 'remains over as the eternal world-eye' after death. I used to think that the phrase, 'the eyes are the windows of the soul', meant we could look deep into the eyes of another and get a glimpse of their essence, but now I wonder if there's an alternative meaning, looking out through the windows rather than into them. Could it be that our body is essentially an avatar and the one consciousness looks out through our bodily eyes in order to experience the differentiated aspects of itself?

We humans, or at least, we Western humans, grow up in a culture that emphasises our individuality. In *Selfie: How the West Became Self-Obsessed*, Will Storr traces this individualistic mindset back to Ancient Greece, a country of islands, inlets and small city states, a geography that shaped its people:

> 'A person's worth, and success in rising up in society, depended largely on their own talents and self-belief. Celebrities were hailed. Beautiful bodies venerated. A particulate landscape became a particulate nation became a particulate people with particulate minds.'[14]

He contrasts this with the Eastern mindset, which evolved amidst the more undulating geography of China:

> 'In contrast to the crags and islands of Greece, most of his [Confucius'] country's population lived on great plains and amongst gentle mountains ... For the Confucian everything in the universe was not separate, but one. It followed from this that they should seek, not individual success, but

> harmony. This perspective has a number of profound implications for the way the East Asian self experiences reality.'[15]

So how can we overcome our fragmented, individualistic Western worldview? I realise I'm quoting Einstein a great deal, but, unsurprisingly for someone whose name is synonymous with genius, he said a lot of interesting things, including: 'The true value of a human being is determined by the measure and the sense in which they have obtained liberation from the self'[16] and 'a new type of thinking is essential if mankind is to survive'.[17]

The Meaning of Life

Our beliefs about who we are, what life is for, and how the world works have very real implications for how we live our lives. What *is* life all about? Does it have meaning? If so, what is it? And how do I as an individual fit into that greater purpose?

There isn't a right or wrong answer to this. Whether life has meaning, and what it may be, is a matter of choice. We can choose to believe that life is fleeting and futile, nasty, brutish and short, and that death is final, but that story doesn't work particularly well for me.

I've chosen to believe that life may not have *intrinsic* meaning, but that I can choose to *give* it meaning, and thereby live purposefully. Over the course of my lifetime, I've tried on several purposes for size, and some options have definitely proved more life-enhancing than others.

For a long time, I was conditioned to focus on achievement. That's what our society rewards and celebrates – the top grade in an exam, the promotion at work, the gold medal.

Then I went through my materialistic phase – lusting after the huge country properties in the glossy pages of *Country Life*, with their designer interiors and beautiful antiques.

If those things work for you, then go for it. I wish you happiness (although you might want to remember that people who define success in terms of external achievement or acquisition tend to be less happy with their lives – just FYI.)

One thing I've learned is that if I catch myself thinking, *I'll be happy when ...* then I almost certainly won't be. If my happiness is conditional on achieving or owning something, then it will be a flimsy and fleeting kind of happiness. The grasping ego-mind will soon be wanting something else, and will once again be promising itself, *I'll be happy when ...*

The guiding principle that has worked best for me is to focus on maximising my personal evolution, in service to the whole. This principle starts from the foundational belief that we are all fragments of the one divine consciousness from which all else derives. Barbara Marx Hubbard would call this the universal evolutionary impulse; the one divine consciousness wanted to embark on its own personal growth journey, and recognised that growth happens through experiences and interactions. Evolution relies on contrast, as we have already seen, at the creative edge between one entity and another. It was rather boring being the one divine consciousness, the only being in the universe. When you're omnipresent and omniscient, there is literally nowhere to go, and nothing to learn. No contrast, no relationships, no surprises, no fun.

So divine consciousness splintered itself into a countless number of fractal consciousnesses, of which you are one, and I am one. This way, it could experience itself from an infinite number of perspectives, living an infinite number of lives, learning an infinite number of lessons. And so here we are, you and me, fragments of the one original consciousness.

Alan Watts puts it this way:

> 'The only real "you" is the one that comes and goes, manifests and withdraws itself eternally in and as every conscious being. For "you" is the universe looking at itself from billions of points of view, points that come and go so that the vision is forever new.'[18]

So, to be in alignment with the divine plan, we're meant to live as rich and juicy a life as we can, to experience everything to the max, so we contribute as much as we can to the evolution of the one divine being that brought us here. Life isn't meant to be too easy and comfortable, or else we're not growing. Greed is good in this context – not greed for material riches, but rather a greed to gorge ourselves on life in all its infinite variety.

Hidden Journey is another of the life-changing books I read during my retreat in Ireland. In his memoir, the religious scholar Andrew Harvey relates the answer given by his spiritual teacher, Ma, when a student asked her, 'What should I ask for?' She replies:

> 'Ask for everything. Everything. Do not stop at peace of mind or purity of heart or surrender. Demand everything. Don't be satisfied with anything less than everything. Our Yoga is the transformation of human life into Divine Life here on earth; it is a hard Yoga, and it demands those who have the courage to demand everything, to bear everything, and to ask for everything.'[19]

Put rather more irreverently by the iconoclastic journalist, Hunter S. Thompson:

> 'Life should not be a journey to the grave with the intention of arriving safely in a pretty and well-preserved body, but rather to skid in broadside in a cloud of smoke, thoroughly used up, totally worn out, and loudly proclaiming "Wow! What a Ride!"'[20]

These beliefs – that life is to be lived large, and the more boldly we live the more we contribute to the evolution of collective consciousness – have had very practical implications for the way I live my life. My perspective changed in an instant when I read Neale Donald Walsch's imperative: 'Do not allow your life to represent *anything* but the grandest version of the greatest vision you *ever had* about Who You Are.'[21] Since then, I've sought out the places on the edge of my comfort zone, or sometimes beyond. My motto for life seems to be: bite off more than you can chew, and chew it anyway.

Pain is Inevitable, Suffering is Optional

From this perspective of desire to live life in all its richness, there's no need to label experiences as *good* or *bad*, to try to hang onto them or rush to get them over with. Everything changes soon enough. In the meantime, we're free to make a choice about how we respond.

I offer the folk tale of the Taoist farmer, as told by Alan Watts:

> 'There was once a farmer in ancient China who owned a horse. "You're so lucky!" his neighbours said, "to have a horse to pull your cart."

"Maybe," the farmer replied. One day he didn't close the gate properly and the horse escaped.

"Oh no! This is terrible news!" his neighbours cried.

"Maybe," the farmer replied. A few days later the horse returned, bringing with it six wild horses.

"How fantastic! You're so lucky," his neighbours told him.

"Maybe," the farmer replied. The following week the farmer's son was breaking in one of the wild horses when it kicked him and broke his leg.

"Oh no!" the neighbours cried, "such bad luck!"

"Maybe," the farmer replied. The next day soldiers came and took away all the young men to fight in the war. The farmer's son was left behind because of his broken leg.

"You're so lucky!" his neighbours cried.

"Maybe," the farmer replied.'[22]

Often, things that seem calamitous at the time turn out to be blessings in seriously heavy disguise. In *What Doesn't Kill Us*, Stephen Joseph describes his experience of working with survivors of the *Herald of Free Enterprise* ferry disaster just outside the port of Zeebrugge in Belgium in 1987, in which 193 people lost their lives. Survivors were left shocked and traumatised by what they'd experienced.

However, some time later, many of the survivors described the experience as life-changing in a tremendously positive way. The benefits fell into three categories: valuing their friends and family more, and feeling more compassion; growing in wisdom and self-acceptance; and feeling more gratitude for the simple things in life, becoming less materialistic and more able to live in the present. He quotes research showing that 30 to 90 per cent of people confronted by horror and adversity are burnished in the crucible of catastrophe into wiser, more fulfilled people with stronger relationships, fresh perspectives and greater resilience.

Stephen Joseph quotes Terry Waite, the Anglican envoy who was taken hostage in Lebanon in 1987. I'd heard Terry speak at an event in Oxford shortly before he was seized, so I felt a personal connection to his fate. He was held in solitary confinement for four years – and unlike a prison sentence, hostages don't know how long they will be held for, or even if they will be released alive. He said of his experience:

> 'Suffering is universal: you attempt to subvert it so that it does not have a destructive, negative effect. You turn it around so that it becomes a creative, positive force.'

Pema Chödrön's wise and lyrical writings on Buddhism have rightly found a wide audience in the West. She reminds us that the first and most fundamental teaching of the Buddha is that humans will suffer as long as we believe in the permanence of things, and count on them for our sense of security, that it is in the nature of things to disintegrate and leave us untethered.[23]

So it's not impermanence that causes suffering; it is our resistance to it. When we accept impermanence – and insecurity – as the way of the world, paradoxically everything gets much easier. One of the most liberating sentences I ever read was from Tristan Jones, the one-legged Welsh sailor: 'The only true riches in life are to be found between your ears.'[24] It allowed me to release my craving for illusions of security, and to trust that I, in myself, contain everything I need, and there is nothing outside of me or in the future that can add to or diminish that fact.

The Uncarved Block

For the first couple of months of the Atlantic voyage, I found myself constantly doubting whether I was doing things the right way. How was ocean rowing supposed to be done?

This is the problem when we grow up in a culture that has deemed a small number of ways *right*, and a whole load of ways *wrong*. Eventually I realised that, given the duration of an ocean crossing, it's more of a temporary lifestyle than a sporting event. It helped me to see it as a microcosm of life, rather than a race. I may not have had much experience of rowing oceans, but I did have considerable experience of life, so I would give up on trying to row the Atlantic the *right* way, and as Frank Sinatra would sing, simply row it *my* way. This worked much better.

Taoism has a word for knowing who we are and living life in accordance with our essential being. *Ziran* means, literally, by-itself-so-ness, or *naturalness* in the sense of being true to its own nature. The metaphor is the

Uncarved Block, or *Pu*. In *The Tao of Pooh*, Benjamin Hoff holds up Winnie the Pooh (Winnie the *Pu*) as the epitome of the Uncarved Block.

> 'The essence of the principle of the Uncarved Block is that things in their original simplicity contain their own natural power, power that is easily spoiled and lost when that simplicity is changed.'[25]

This ties in with the idea of heterogeneity that we previously encountered in chaos theory. Individuals and collectives thrive when people can be themselves, rather than shoehorned into a societal template of how they should be.

In *The Gene Keys*, Richard Rudd emphasises the connection between cherishing our individuality and the wellbeing of society as a whole, positing that each of us can contribute to the evolution of society only when we have done the inner work of discovering our own identity and unique gifts. The beauty of our individuation, as he sees it, is that the more differentiated we become, the better we are able to harmonise with the other members of our society.[26]

Living elegantly is associated with acting spontaneously, optimally attuned to the natural flow of things, because this requires the least energy and effort so is maximally sustainable. This isn't to say that nothing will ever change; what is natural at one stage of life may not be natural at the next, so *ziran* is entirely compatible with growth and evolution.

This process tallies with the archetypal journey set out by Carl Jung in his theory of individuation, which begins with the individual asking: *Who am I? Why am I here?* He acknowledges that this work is not easy.

> 'It is a most painful procedure to tear off those veils, but each step forward in psychological development means just that, the tearing off of a new veil. We are like onions with many skins, and we have to peel ourselves again and again in order to get at the real core.'[27]

I went through this process on the Atlantic, and it was indeed painful. On Day 49 of the voyage, I wrote in my blog post:

> 'When I told my mother about the latest casualties [more equipment broken and lost] she commented, "The ocean is really stripping you down, isn't it?" And this is true, metaphorically as well as literally. As I'm

> left with less and less, it makes me realise how little I actually need, how little is actually important.'[28]

As well as my boat, my sense of self was also getting stripped down, to its barest essence. Extended solitude will do that for you. In everyday life, we're unconsciously playing roles appropriate to the context, revealing different facets of ourselves depending on whether we're with colleagues, boss, friends, parents, children, or spouse or partner. In 'Letter to a Young German', written in 1919, Hermann Hesse describes the importance of seeing beyond these roles:

> 'You must unlearn the habit of being someone else or nothing at all, of imitating the voices of others and mistaking the faces of others for your own.'[29]

Alone on the ocean, I had no need to play any role. When my satellite phone broke, severing all communications with shore, I was in relationship with nobody and nothing apart from myself and my surroundings, completely immersed in the experience. My solitude was complete, and I had to take total responsibility for my state of mind, for better or worse. The state of my consciousness became the primary, almost the only, determinant of my experience. John Milton's words say it perfectly:

> 'The mind is its own place, and in itself can make a heaven of hell, a hell of heaven.'[30]

Despite my mind often going for the hell option, overall I found my solitude liberating. As Carl Jung said, I peeled back the layers of my identity, letting go of identification with my gender, nationality, ethnicity, education, profession, letting go of my constructed self. Like many ocean rowers, I preferred to avoid the encumbrance of salty, sweaty clothes so I usually went naked at sea, adding to the sense of paring back to the bare essentials, with no armour to protect me, no clothes to make a statement about who I was.

I wondered what I would find at my core: who or what would I be when all was stripped away?

My answer is: I found nothing. Maybe there were more layers still to go; I thought I was stripped down as far as it was possible to be, but at no point did I completely lose the sense of *I*, although at times the *I-ness* was little more than the awareness of being a sensing creature experiencing the warmth of the sun on my skin, the sound of the wind in my ears, the rocking of the boat. I'd wondered if I might find, buried deep in my core, an essential self, a luminous soul that I could point at and say with certainty, *I am that*, but it never happened. I choose to interpret my lack of a *me* as supporting Neale Donald Walsch's view that we are all fragments of a unified consciousness, so there was no *me* because, at that deep level of being, we are all one. The failure of my search for my essential self is echoed in Richard Rudd's words:

> 'The more you look for your own true identity, the more ephemeral it becomes ... you cannot be who you are as long as you think you are someone, but you still have to set off and search for this someone in order to realise that they do not exist.'[31]

This chimes with Tenzin Palmo's words:

> 'That was the Buddha's great understanding – to realize that the further back we go, the more open and empty the quality of our consciousness becomes. Instead of finding some solid little eternal entity, which is "I", we get back to this vast spacious mind which is interconnected with all living beings. In this space you have to ask, where is the "I", and where is the "other"? As long as we are in the realm of duality, there is "I" and "other". This is our basic delusion – it's what causes all our problems.'[32]

When we look at Hoffman, Kastrup, Taylor and a gaggle of world-renowned physicists on one side, and Palmo, Chödrön, Rudd, Huxley, Watts, Walsch, Winnie the Pooh, *et al.* on the other side, it appears that science is converging with ancient wisdom from just about every religious tradition across the world. I imagine science and spirituality divorcing each other around the time of Galileo and the Age of Enlightenment in the seventeenth century, and after several hundred years of estrangement, they have come full circle and are now starting to flirt with each other again. We may yet live to see them consummate their reunion.

As the joke goes, when the scientists finally make it to the mountaintop, they find a band of raggedy monks who have been sitting there for centuries. The monks say, 'What took you so long?'

If ignorance of our nature is what keeps the wheel of *samsara* turning, and if we overcome that ignorance to understand ourselves as a fractal of a universal consciousness, and death to be merely a transition from one form of consciousness to another ... can we stop the wheel?

PART FOUR

THE NEW PARADIGM

'On the other side of a storm is the strength that comes from having navigated through it. Raise your sail and begin.'

Commonly attributed to Gregory S. Williams

Chapter 13

Preamble

'All that you touch, you change. All that you change, changes you.'[1]

Octavia E. Butler

The day my life changed was a dull, grey January day in the late 1990s. It started much like any other – the alarm went off at 6 a.m., blaring at me to get up and be a grown-up. I went through the morning ritual of showering, donning my young professional outfit like a suit of corporate armour, and boarding the commuter train, newspaper in hand, to trundle through the dark and drizzle into the City of London for another dreary, desk-bound day.

But as I sat on that train, gazing absently at my fellow passengers, similarly besuited and pasty-faced from lack of sunshine, a feeling that had been simmering in my subconscious slowly rose into awareness like a rock emerging from an ebbing tide.

I couldn't keep doing this. This professional life seemed just fine for most folks, certainly most folks I knew. My friends, and obviously my colleagues, appeared to accept this corporate, commuting charade as a normal way for adults to live. But I was miserable. I felt like a sickly seedling, deprived of daylight, meandering around phototropically, trying in vain to find a sunbeam where I could thrive. Something had to change. *I* had to change something. I felt, deep down, I had something to offer. I had no idea what it might be, but I was pretty damn sure I wasn't going to find it in management consulting.

That night, I sat down and wrote two versions of my own obituary – the one I wanted, and the one I was heading for if I carried on as I was. I'd tried before to figure out what I wanted to be when I grew up, but the advice I'd gleaned from books always seemed to start with creating an inventory of existing skills and experiences. Apart from this list being depressingly short, it also pointed inevitably to doing more of what I was already doing. It was a linear, limited and limiting perspective, a low vantage point to consider what might be possible, feasible, practicable, sensible. What I wanted, *needed*, was radical change.

Now, pen in hand, I completely stepped out of the present and took a quantum leap through time to the end of my life. I left my limited self behind and dared to dream of all that I could become between now and my final breath. Suddenly, I felt a rush of excitement. My horizons expanded exponentially. I could see all I hoped to accomplish, the life I hoped to have created. I could choose from a dazzling array of options, and anything seemed possible. The vision of my future self that surged through me that day seemed to come from another realm, a parallel universe in which I was living the life I was born to live, far removed from the life I was actually living. It was as if I'd asked my soul what it wanted to become in this lifetime, and it had cried out joyfully, 'I thought you'd never ask!' and presented me with a vision that was inspiring, fulfilling ... and seemingly utterly unattainable.

And yet, somehow, I've attained it. Scary as the vision was, once I'd seen it I couldn't un-see it. Subconsciously more than consciously, by accident more than design, the ship of my life gradually turned in the direction I wanted to go. It was a messy process, full of missteps and even apparent disasters, but it instilled in me the trust that there is method in the universe's madness, that we don't need to know the way before we start the journey, and that the messes might even be essential stages on the way to the promised land.

After all, birth is a messy business. Why should rebirth be any different?

Just as crisis precedes transformation, stuckness precedes breakthrough. During those latter years of my so-called career, I felt trapped, cornered. I couldn't see a way out. I was living in a self-reflecting yuppie bubble and the world outside the bubble was invisible to me. I was caught in the realm of what I knew I knew, and until I burst the bubble I didn't know what I didn't know about the infinite possibilities that lay beyond.

But breakthroughs, alas, are not one-and-you're-done. Life – or at least, *my* life – has been a repeating cycle of stuckness, frustration, even more frustration, unbearable frustration, and then the heady *a-ha!* of insight.

Writing this book has been much the same. Here, I'm going to do the literary equivalent of breaking the fourth wall, and interrupt my own narrative to share something of the process that I went through in writing it – or, for a while, failed to write it.

I'd written three books already, and this one was to be based on my doctoral dissertation, which lulled me into a false sense of security. I thought I already knew what I needed to know, and all I needed to do was translate the dissertation out of academic language and into something that normal people might actually want to read.

Parts One to Three flowed relatively easily. I was on a tight deadline from the publisher, but it looked like I would have no problem achieving it. I thought I was well on the way across this literary ocean, but I was about to discover I hadn't even made it out of the harbour.

This became apparent when I ran slap-bang into the brick wall of Part Four. This section was supposed to be a bold vision of a beautiful future, a utopia of milk and honey, justice and prosperity for all. I'd assumed I knew what a better future would look like. After all, I'd spent the best part of the last two decades obsessing over it.

But my mind's eye was blank. I had no problem identifying what I *didn't* want. I just didn't know what I *did* want.

Part Four became my nemesis, the brick wall impenetrable. I tried meditating, drawing oracle cards, free association writing exercises, breathwork, procrastination, discipline. Nothing helped to unstick my stuckness. I was all too aware of the looming deadline for the completed manuscript, just weeks away, and I had absolutely no idea how I was going to pull together all the strands I'd ambitiously set up in the first three parts. I felt like the writer of a murder mystery who has written most of the book, but still hasn't figured out whodunnit.

It gradually dawned on me that something needed to shift *in me* before I could find a way around, over, or under the wall. The problem, I now realised, was that for all my brave talk about change, I didn't particularly want to change myself. I was doing okay as I was, and change is hard work.

In 2005, when I set out on the Atlantic, I was intentionally setting out to reinvent myself as an adventurer. My old life had very obviously not been working for me – the endless pursuit of material wealth had left me feeling lost, confused, worthless. I left my old life with barely a backwards glance.

This time around, life was pretty much fine. I'd built a steady career in corporate speaking, which gave me opportunities to travel (at least until Covid), meet new people and make a positive difference, while leaving me enough spare time to pursue passion projects. I was happier in myself than I'd ever been. From the ashes of my city career had arisen a new version of me that I rather liked, even respected. I had a circle of wonderful friends, felt purposeful and fulfilled, and was grateful for all the blessings in my life. I didn't feel the desperate need to reinvent myself that I had twenty years ago.

But heck, what does the universe and its relentless evolutionary impulse care about my ego-based needs and wants, my smugness and self-satisfaction, my pretensions and privilege? I should know by now that it cares very little. Change was coming, whether I liked it or not.

Even in the midst of my confusion and frustration, I could see the irony and perfection in my stuck state. I'd become a microcosm of the bigger question I was trying to answer – how does change happen, and why does it fail to? The gift in my writer's block was that I was experiencing individually what we're going through collectively. I was embodying the very problem that I was describing, and was feeling the full force of the ego's fear at letting go of the tried, tested and trusted approaches that have worked so well in the past.

It's easier to change when the old way is clearly failing – when an individual is miserable, or when a society is collapsing. It's harder to change when the old way still works just about well enough – not great, maybe, but if it ain't broke, don't fix it. We can interpret *ain't broke* very generously when we expect change to entail considerable effort and inconvenience.

I was feeling the fear of releasing outdated strategies and dying identities, even though I could see they were constricting me, and I needed to shed them as a snake sheds its outgrown skin. If I was going to ask anybody else to go through this fear in order to embrace a brighter, braver future, it was only right that I should be willing to blaze that trail.

In short, my strategies, tried and tested and refined for the last fifty-four years, were no longer any use to me. I ruefully recalled Margaret Wheatley's words – about experiencing 'so much frustration that we surrender. Only then are we humble enough and tired enough to open ourselves to entirely new solutions.'[2]

At last, one night at around 3 a.m., as I lay in bed not sleeping, inspiration struck. I would have preferred it to strike at a more sociable hour, but I was willing to take it whenever I could get it, and maybe inspiration only gets its chance when the conscious mind drops its guard – in the in-between moments, the liminal spaces, the dark corners of the night.

I realised a variation on the obituary exercise could help me once again: to project myself into the future and dare to dream of the world as I would like it to be. I needed to write from my heart, not head; from soul, not from cerebellum, to create a vision of a world of equal opportunity, harmony and thriving. If my past experience was anything to go by, painting a vivid vision of the future is enough to start the shift. The vision becomes the destination, and as the journey unfolds it becomes clear which forks in the road bring us closer, and which take us further away. We have a north star by which to navigate. There may well be a few diversions and dead ends, but ultimately even the dead ends become an important part of the process. We simply need to keep the faith, and hold true to the vision.

Once again, just as with the obituary exercise, as soon as I opened that portal into the future a whole new world of possibility flooded into my mind's eye. I could see a thriving new civilisation in luminous detail. I could feel the joy of the new ways of living and being. I had an intuitive sense of the principles that would undergird this New Paradigm.

I also sensed that things would have to get worse – a lot worse – before they could get better, that the entrenched power structures won't willingly surrender their influence or share their affluence. But wresting power from the current elites wouldn't be the answer. Humanity as a whole is wedded to the status quo, no matter where in the pecking order we currently rank. Even at the same time as recognising its shortcomings, we also find the current regime reassuringly familiar, flaws and all. It's hard to think outside the box when the box represents structure and security.

We might deny that we're in a box at all, but we are – a box of habits, narratives, systems, beliefs, norms, conventions and cultural conditioning that prevents us from ripping it all up and starting again from first principles.

So, in the story that follows, I'm going to blow up the box.

Is it possible to generate positive change without exploding the current model? Given the literature on collapse, it seems unlikely. Certainly, change can and does happen gradually – what adrienne maree brown calls *slides* rather than *shocks* – but given the speed and scale of our looming existential crisis, will incremental change be sufficiently radical? The problem with gradual change is it doesn't send a recognisable signal that the rules of the game have fundamentally changed. It's hard to recognise a paradigm shift when you're in the middle of it. I'm not a fan of Milton Friedman's economics, but it's hard to disagree with his observation on how change happens.

> 'There is enormous inertia – a tyranny of the status quo – in private and especially governmental arrangements. Only a crisis – actual or perceived – produces real change.'[3]

The vision that I'm about to share may seem unnecessarily bleak to begin, but it's intended to be taken metaphorically, to liberate our imaginations. I suggest you read the following two chapters as a thought experiment, a fictional narrative in which humanity gets to wipe the slate clean and start over, bringing our best wisdom to bear as we design a new chapter for humanity – a world interrupted, a world reimagined.

Kali the Destroyer of Worlds

Transport yourself forward through a wormhole in the space-time continuum, to a year seven generations hence, 2197. Imagine that sometime during the twenty-first century, there is an enormous solar flare – let's call it the Kali Flare, in honour of the Hindu goddess of destruction and rebirth – that instantaneously knocks out all electronic devices and power grids across the whole planet. Anything that runs on electricity is rendered inoperable, useless, kaput.

(This is a disturbingly plausible scenario. A coronal mass ejection caused the Carrington Event of 1859, the most intense geomagnetic storm in recorded history. Electricity didn't become a feature of human civilisation until the 1880s, so Carrington's impacts were limited to a particularly spectacular display of the aurora borealis and a few fires in telegraph stations. A similar event now would cause widespread blackouts and catastrophic damage to electrical power grids. Carrington was graded G5, the most extreme grading, yet it pales into insignificance next to the Miyake Event in 774 CE, which produced the largest and fastest rise in carbon-14 ever recorded, fourteen times that of Carrington.[4] There will certainly be another significant solar flare at some point. We just don't know when.)

The Kali Flare doesn't hit Earth itself. The planet isn't engulfed in a fireball. What causes the damage is an enormous magnetic pulse, carried to Earth by solar winds, creating a geomagnetic storm that lasts for days. What has taken centuries of development to create is destroyed in an instant. All digital data is annihilated, along with all the equipment that stores, transmits and manipulates it. Satellites, computers, servers and mobile phones are bricked, as is anything else that requires an electrical current – cars, trucks, refrigerators, heating and air conditioning systems, industrial plant, hospital equipment, agricultural machinery, streetlights, house lights ... *everything*.

Before the flare, human civilisation is already teetering on the brink of collapse – temperatures rising, ice caps melting, wildfires rampaging, volcanoes erupting, earthquakes shaking Gaia's tectonic plates. Kali delivers the existential *coup de grâce*. Earth effectively returns to the Dark Ages.

This is catastrophic for humanity. Many die in the immediate panic and chaos, but most in the long, painful aftermath. Supply chains are abruptly and irreparably broken. Without electric pumps, domestic water supplies fail. There is widespread famine and disease. Humanity can't even begin the process of reconstruction, because most industrial tools depend on electricity. We can't bootstrap our way up, because Kali has taken away our bootstraps. For a supposedly intelligent species, we've made ourselves incredibly vulnerable to a single point of failure. We are, literally, *powerless* in the face of cataclysm.

For billions of people, it feels as if the world is ending – and in a sense, it is. The suffering is tremendous. Humanity is brought to its knees. Yet, strangely, there is also a prevailing acceptance of our fate, even amidst the devastation, loss and grief. It is as if, on some level, we know that we were never going to change unless and until we were forced to. Somehow, subconsciously, we've been longing for this catharsis, to stop the insanity that our civilisation has become, the opportunity to start over and create a better, fairer society that works for everybody.

Kali is a great leveller. Money is revealed for the fiction that it is, nothing more than digits in a computerised bank account. Suddenly there are no computers and no banks, so effectively no money. It's a shock to the rich and privileged to discover that all their wealth cannot save them. Their money is worthless, and they don't have the resourcefulness, resilience and skills that life has required of the less privileged.

Rather than a leveller, in fact, it might be more accurate to say that Kali tips the seesaw. The greater the privilege, the greater the dependence on modern technological infrastructure. The wealthy suddenly find themselves as helpless as babes, while those in slums and shacks, in jungles and on remote islands, the hunter-gatherers and uncontacted tribes, barely notice the difference. (Most people reading this book – and certainly the person writing it – would be at the 'helpless' end of the spectrum. I'm an accomplished forager in the aisles of Waitrose, but not much beyond.)

The survivors are mostly those regarded as the oddballs of society, the ones who had already been questioning the dominant paradigm of the twenty-first century. They are the mavericks, rebels and dropouts. Also, indigenous peoples, so long treated as second-class citizens, are well prepared in practical terms, and especially they are spiritually prepared. With their cyclical cosmology, they embrace the cataclysm as a turning of the wheel, the necessary darkness before the dawn of a new era. The indigenous peoples and the independent thinkers band together, and call themselves the Indies.

These groups have often felt marginalised and looked down upon, but now they come into their own. Their consciousness was already shifting, and now it seeds the next era of the human story. The balance of power swings from those who benefited from the old way to those who were disenfranchised and disempowered by it. The people who are all too aware of the shortcomings of the old system are the ones who begin to design a new system that works for all. The meek – and the eco-geeks – finally do inherit the Earth.

Chapter 14

Principles

'What if this darkness is not the darkness of the tomb, but the darkness of the womb?'[1]

Valerie Kaur

Imagine that you are one of these Indies, emerging from the wreckage of the old to design a new way for humanity to inhabit the Earth, a new guiding narrative by which to live. What values might inform your vision? What guiding principles might scaffold the new civilisation? What story might support what Alex Evans in *The Myth Gap* calls 'a larger us, a longer now, and a better good life'?[2]

I offer some ideas here – you're welcome to try them on for size, and see how they resonate with you. They may serve as a useful starting point, which can only be proven one way or another by adopting them, adapting them and refining them as they transition from abstract imaginings into embodied, experienced ways of being.

We don't need to wait for the world to collapse before we start experimenting with these concepts. You might decide that you'd like to start living a happier, healthier, more fulfilling, regenerative and resilient life right now. Why not start now?

Coevolution

What is the point of being human? What is our purpose?

An impartial observer watching the people of the pre-Kali world might conclude that our primary goal was the acquisition of stuff, for which we first had to acquire money. Large portions of the world didn't have enough money to live comfortable lives, or even to eat, so they were preoccupied with trying to meet their basic needs. This wasn't their fault – it was a fault in the system. Many of them worked incredibly hard, leaving little time for important human activities like creativity, learning or spending quality time with family. But even once people had enough to live comfortably, often they continued to work many, many hours a week, selling their precious and irreplaceable time to large organisations so they could acquire more material things.

From the perspective of 2197, this looks like a kind of mass psychosis, a competition to see who can acquire the most stuff in one lifetime, always comparing your possessions against the next guy's, even though psychologists kept publishing more research to show there was little correlation between wealth and wellbeing, so gauging the success of a life in terms of material wealth seems a really poor idea. Meanwhile, materialism led to terrible environmental problems. People were willing to exploit and pollute natural resources with no regard for the future, all in service to this obsession with inanimate objects.

Many, if not most, humans seem to have an innate drive to want more; if some is good, then more must be better. So the question becomes: what do we want more *of*? What if the new civilisation reorients from more matter to wanting more personal and spiritual growth, more fulfilment and joy? What if evolution of self, and the coevolution of society, become the driving forces?

From this perspective, humanity could develop a fundamentally different relationship with possessions. Material goods would be relegated to a lower status, necessary for satisfying the basic needs in Maslow's hierarchy,[3] but no longer regarded as the measure of somebody's worth. We would aspire to elegant simplicity, appreciate enough-ness, while conspicuous overconsumption would be regarded as an attempt to make up for feelings of spiritual disconnection and personal dissatisfaction. In the IPAT equation, the desire for Affluence would largely dissipate in favour of Adequacy.

Of course, not everyone has to (or ever will) share the same beliefs. If people wish to organise their lives around stuff, that's their prerogative. The new civilisation embraces the principle of non-violation – everybody is free to live however they wish, provided that it doesn't impinge on the freedom of others (including future generations) to live as they wish. In this new paradigm, materialists might be regarded as oddballs and eccentrics, because it's such an obviously perverse lifestyle choice, but live and let live.

What does a civilisation look like when the evolution of consciousness is the main goal? How do we embody that as a guiding principle? How do we know how to live?

Indigenous culture offers inspiration. Ilarion Merculieff eloquently describes how the Unangans of the Arctic tune into inner wisdom:

> 'When Unangan Elders speak of the "heart", they do not mean mere feelings, even positive and compassionate ones, but of a deeper portal of profound interconnectedness and awareness that exists between humans and all living things. Centring oneself there results in humble, wise, connected ways of being and acting in the world. To access it, you must "drop out" of the relentless thinking that typically occupies the Western mind. Indigenous peoples have cultivated access to this heart source as part of a deep experience and awareness of the profound interdependence between the natural and human worlds. If heeded, this portal provides the inner information that keep us in "right relationship" with all of life, thus ensuring our long-term survival and wellbeing, individually and collectively. When guidance or information comes from the heart, it can be relied upon and has impeccable integrity, whereas our fallible thought processes regularly deceive us.'[4]

This is a powerful state of awareness, getting the fallible ego-mind out of the way to access the realm where humans are in tune with each other, other living beings and the Earth herself. The guidance that comes from this place is pure, creative and compassionate, bypassing the biases and fallacies that plague our cognitive minds.

This solves the apparently inescapable conundrum of the G.I. Joe fallacy. If we try to out-think our thinking processes, we are trying to solve the problem from, quite literally, the level of consciousness that created it. Thinking back to Einstein's dictum about our society having forgotten that the intuitive mind is a sacred gift, the Unangan approach puts heart and mind back in their rightful places.

What contributes most to our evolution is anything that hasn't been done before – a new experience, an interaction with a new person, creating a new piece of art or music or literature. Humans love novelty, and in the materialistic culture this was hijacked by the advertising industry into persuading people to buy new stuff. But stuff was a poor substitute for truly enriching novel experiences. This is why the body releases the happy hormones – serotonin, dopamine, oxytocin and endorphins – in response to novelty, because we've been designed to explore, to grow, to evolve. Our physiology rewards us for trying new things, embracing fresh adventures.

The old culture of fear, perpetuated and exploited by political, media and corporate interests, squished our natural tendency to adventure. Combine this fear-filled messaging with our old materialist perception of ourselves as individual, mortal souls, and it's small wonder we were held captive by our fears. Failing to see ourselves as eternal beings united in a field of collective consciousness, it's understandable that we felt isolated, lonely, and afraid of death. What poor, constricted lives we led, held inside our prisons of fear, prisons that might keep dangers out, but also kept us in.

This isn't to advocate recklessness. Rather, it emphasises the *quality* of life, rather than the *quantity*. A hundred years of playing it safe contributes a lot less to the collective consciousness than forty or fifty years of living boldly. The soul has nothing to fear from death. It's only the ego that does.

The human animal created the ego for a reason. The ego is the small self, housed in a mortal body, that needs to keep itself alive because the ego, unlike the soul, dies when the body dies. So the ego's job is to do whatever has to be done to meet its bodily needs, to be vigilant for threats to its continued existence, and to get combative when it encounters a perceived enemy. The ego is the temporary suit of armour that we put on to get us through this lifetime, to help us navigate the Earth realm, to create the boundary between our soul and the other souls in order to create the contrast and interaction necessary to the evolutionary process. The ego thinks this body and this personality is all it is, and that death is the end – and it's right, because death *is* the end for the body and the ego.

But death is not the end for *you*. The soul is the eternal part of us that has been incarnate in the past, and will be reincarnated in the future. It's the part that knows the soul contract it signed up for, and accepts death as a return to home. From the new level of consciousness, we know that after death our essential soul lives on – as does our contribution to the collective evolution of humanity.

Connection

The prevailing worldview in the early twenty-first century is that we are isolated individuals, separate from each other and from our Earth, but we are increasingly starting to understand that what we think separates us actually connects us.

From inside our own minds, the whirlpool of our individual consciousness makes us believe that we're separate from other people. I know what's in my mind but I don't know what's in your mind, so I feel alone in here. But as we've already seen, there's growing evidence that consciousness is not necessarily bound to the brain, it is all around us, with the brain as receiver and filter.

An oceanic analogy: from the perspective of land, we think the oceans separate us, country from country. But from the perspective of a seafarer, the oceans *connect* us. The difference is having a vessel that allows us to escape our limitations as a land-dwelling animal and to journey from one land mass to another. This vessel could be meditation, psychedelics, psychic abilities, a near-death experience (don't try this at home) or simply our own imagination – any process that enables us to transcend our physical bodies and escape the jail of our minds to voyage across the seas of consciousness. We realise that there are no boundaries, no separation, only connection.

Another analogy would be the mycorrhizal network: the fine network of fungal fibres that weaves a web between the roots of trees, distributing nutrients and even sending alarm signals of insect invasions. We normally don't see the mycorrhizae, hidden beneath the forest floor, but they're crucial to the health of the trees as individuals and the forest as a whole, a vast communication matrix pulsing with information and nourishment.

Martin Luther King Jr knew this:

> 'All life is interrelated. We are all caught in an inescapable network of mutuality, tied in a single garment of destiny. Whatever affects one directly, affects all indirectly. We are made to live together because of the interrelated structure of reality.'[5]

In the new civilisation, rampant individualism gives way to the paradigm of connection. People, animals, plants and planets all exist in one cosmic stream of consciousness as differentiated but complementary and connected aspects of the whole. We reconnect with the ancient animistic knowledge of our interdependence, not just with other humans, but with the whole Earth, indeed, the whole cosmos. In this worldview, there is no 'away', as in throwing something *away*, and there is no 'them', as in *us and them*.

In conjunction with the principle of coevolution, the guiding question becomes: how can I use this day to contribute the most to evolution – my individual evolution and the evolution of the collective? In effect, these are the same thing; all individual consciousnesses are connected within the species-wide morphogenetic field, so anything that benefits the individual consciousness contributes to the collective. If we are all fractals of the collective consciousness, with a shared purpose to evolve, our interactions with the world around us take on a different flavour. Every action and every choice become opportunities for new and educational experiences, whirling with our fellow humans in a dance of coevolution. A rising tide of consciousness lifts all boats (while a rising tide of wealth sadly doesn't).

When Einstein said that problems can't be solved from the same level of consciousness that created them, maybe the shift into this deep sense of connection is what he meant. We've been trying to solve problems from the individualistic perspective, believing that we are one person acting on one problem, or maybe several persons acting on several problems, but still seeing both the problem-solvers and the problems being acted upon as separate things. The current paradigm for addressing social problems tends to go like this: notice that there's something wrong in the world; develop a passionate abhorrence for it; campaign against it. Repeat.

But metaphysically speaking, the more you oppose something, the stronger you make it, because you're adding your energy to it. More than five decades of protesting against ecological damage and social injustices have secured some legal victories in terms of environmental protections and equal rights, but the underlying problems still exist because we haven't transcended the paradigm that gave rise to them. Too often, we're still trying to solve the problem from a consciousness of separation and opposition, when we need to rise to a consciousness of connection and collaboration if we're to achieve a long and lasting solution.

Most humans don't relish conflict, but we've been conditioned to believe this is how change happens – by one party asserting its dominance. But this just substitutes one dominator with another. We need to shift from a dominator mindset to a partnership paradigm.

I'm not suggesting wilful blindness, the hope that if you ignore a problem for long enough it will go away. I'm also not proposing that we stop pressing for political and legal change through nonviolent means. Rather, I'm saying that we need to generate transformation on the metaphysical plane as well as the material, that everything exists in consciousness before it shows up in physical reality, and we know that change happens in the unmanifest realm of thought, intuition and heart-wisdom before it manifests in the material world. As Tenzin Palmo says, all our problems are caused by our delusion of duality,[6] and that same delusion also stops us from solving them. As soon as we admit that we've been propping up the problem, we start to make a difference, because now it's our own conscious awareness we're changing, which ultimately is the only thing we ever can change.

The image that comes into my mind is of the lotus flower. It starts life as a seed in the silt at the bottom of a pond or slow-moving river, down in the sludge of old, rotting, composting vegetation. It emerges from the murky water into the light, where it blossoms into a pristine flower. It transmutes muck into beauty and fragrance.

Another feature of the lotus is that its seeds can lie dormant in the mud, waiting for a favourable time before bursting into life. The oldest recorded seed to germinate was 1,300 years old.[7] We might not live to see the flowering of the seeds of change we sow, but they can bide their time, waiting for more auspicious conditions. We have to sow them anyway.

Neale Donald Walsch says, 'Every act is an act of self-definition. Everything you, think, say and do declares, "This is Who I Am".'[8] And the act of creation extends further than our own selves – in a complex system where everything is connected, everything we think, say and do is also an act of creation within the wider world. No action, even if unobserved or unmeasurable, is ever wasted. We are not just a drop in the ocean, we are the entire ocean in a drop.

Collaboration

At the level of immaterial consciousness, we are already collaborating in creating our future, even if we're not aware that we're doing so. *Conscious collaboration* is the capacity to leverage social interactions and shared endeavours to support our coevolution by pushing our boundaries, and strengthen our sense of connection by aligning around shared goals.

As the work of Frederic Laloux and others has shown, flat organisations that support collaboration make staff feel more empowered and engaged, produce better goods and services and deliver greater customer satisfaction. But too often, collaborative groups in the current paradigm get shipwrecked on the rocks of ego, domination and not knowing how to disagree well.

Effective collaboration becomes much more likely when we shift from identifying with the ego to identifying with the soul. This may sound like a subtle shift but it's radical.

Let's consider the ego again. Because it sees itself as a finite being, the ego tends to get a bit uppity when it thinks it's been insulted. It has a kind of Napoleon complex – it feels small and inadequate, so it overcompensates by puffing out its chest and insisting on its own importance.

The soul, by contrast, has nothing to prove. Whether it's a young soul or an old, it knows it's on an evolutionary journey, and doing the best it can in its current incarnation. It knows its eternal nature, its greatness, its vastness, and it recognises those same qualities in the souls around it. When we see the world through the eyes of the ego, those who disagree with us seem like enemies, and we lapse into aggression, defensiveness, victimhood or self-pity. When we see the world through the eyes of the soul, we see that life is always happening *for* us, not *to* us, and we encounter challenging personalities so they can catalyse our soul's evolution.

I notice that when I'm feeling upset, offended, frustrated, disappointed or angry, asking myself the simple question, 'Why is my ego feeling this way?' is transformational. It reminds me that the emotion, which just a moment earlier had seemed so all-consuming, is actually a choice I've made, and it flows from an earlier choice to identify with my ego. When I shift my perspective from ego-subjective to soul-objective, I see the situation entirely differently. The big angry red balloon of my emotion, full of all my self-righteous indignation, suddenly deflates with an anticlimactic phut.

Imagine how different a group dynamic might be if it was founded on questions such as:

> Why have these particular souls felt called to work together on this project?
>
> What if my ego let go of the need to be right?
>
> How can I embrace 'and' rather than 'or'?
>
> What can I learn from these interpersonal tensions?
>
> How does diversity make us stronger?

What unique contribution does each of us bring?
Where is our optimally creative edge of chaos?

The pendulum of history tends to swing between individualism and collectivism,[9] and after the fiercely egocentric era of the late twentieth and early twenty-first centuries, the swing back to the collective is overdue. The world is too complex, the issues too intractable, for one person to tackle on their own. No matter how heroic or intelligent they may be, they still have blind spots, cognitive biases and cultural conditioning that lead to a one-dimensional perspective. For a solution to work for a whole society, it needs to include representatives of the whole society.

In the future (and already happening in the present), groups founded on trust, mutual respect and psychological safety will be the pioneers. Skills in conflict resolution, nonviolent communication and creative collaboration will be as widespread as basic literacy and numeracy are now. The Hopi Elders say, 'The time of the lone wolf is over. Gather yourselves!' And gather we must.

Compassion

Coevolution, connection and collaboration are important in the new civilisation, but compassion is the most important of all. Compassion is another way of saying love, but not love as a sappy, soppy or sentimental feeling like in the songs – quite the opposite: it is powerful, deep, inspiring, and often very challenging, love interpreted less as an emotion, more as a metaphysical worldview.

Before we can have compassion for others, we have to start by having compassion for ourselves. We can't give what we don't have. We can't accept others if we don't accept ourselves.

The world that we create is an outward projection of what we're feeling on the inside – and to look at the world we're creating now, we have to wonder if we as a species fundamentally dislike ourselves. There is a lot of unhappiness and self-loathing going on, to the extent that we're committing a slow collective suicide. It used to be said that *love of money is the root of all evil*, but it could equally be said, *lack of self-compassion is the root of all evil.* When we feel contempt towards ourselves, we act contemptuously. When we feel compassion towards ourselves, we act compassionately.

Most of us yearn to be seen, to be loved exactly as we are, but at the same time we're scared to be seen in case others don't like what they see. We hide ourselves behind a safe persona that is perfectly serviceable, seems close enough to our real self that it doesn't feel like an out-and-out lie, but it's actually a mask for the parts of ourselves that we fear might be less lovable, not good enough, that might make us less popular, less accepted.

Some of us have pushed those parts down deep, really deep, where they are invisible even to ourselves. We might feel repulsed by people who embody those aspects of us that we don't want to face up to. We want to push them away, because they remind us of the unexpressed and inexpressible pieces of our own self. Maybe on some level we even envy them because they *aren't* afraid of those shadows, but flaunt them for all to see. They are who they are, without apology, and that irritates the heck out of us because we've spent all these years and expended all that energy to suppress the parts we don't want to own. How *dare* they simply be themselves – the good, the bad and the ugly? How *dare* they not fight and sweat and contort and labour to conceal those parts that we've decided are unworthy and must be hidden if we are to be loved?

Trying to be somebody other than who we are is exhausting. It's a heavy cognitive overhead to be constantly editing our words and behaviour. Suppressing this part, pretending that part, wondering what we would say now if we actually were that perfect alter ego we wish to be, carefully cultivating our acceptable public-facing personality ... It's hard work, and all the time and energy we spend pretending is time and energy we don't have to be attentive, thoughtful, loving and light-hearted. It's the difference between being self-conscious, in which we have to be always vigilant in case we accidentally reveal ourselves, and being self-aware, in which we know, accept and love ourselves.

You may well know people who are self-conscious, always checking themselves in the mirror, metaphorically if not literally: how do I look? How do people see me? What do they think of me? Should I say this, do that, sit this way, smile that way? How do I show my best side? In conversation, it feels like you're not getting their full attention, because half of it is on themselves while they wonder if they're doing a good job of pretending to listen to what you're saying.

Hopefully you also know people who are self-aware. They are relaxed, comfortable in their skin, at ease with who they are. They have integrity, congruence and what you see is what you get. These people aren't perfect

– they sometimes screw up, and when they realise it, they apologise and try not to do it again. They don't have to keep up the pretence of perfection in order to preserve some constructed persona. They don't have an overinflated ego that they have to protect in case somebody bursts it.

They operate from their heart, their connection to soul. If all we see of ourselves is our ego, with all its flaws, faults and foibles, small wonder we find ourselves hard to love. But when we see our soul, in all its beauty and perfection, we know that we're not only worthy of love, but we are indeed loved. It's a lot easier to love our expansive soul, with its generosity, compassion, connection and contentment, than it is to love the self-serving ego, with its fears, petty jealousies, wants, desires and narrow-mindedness.

When we identify with soul, we see our work on Earth as sacred, life as sacred, ourselves as sacred. We know we're here to evolve, committed to doing our best. In this lifetime that best might be spectacular, or it may not, and that's fine. There's always next time. We believe we are worth saving.

Compassion fuels our capacity for evolution and transformation. It generates wisdom, peace and harmony, and prompts random acts of kindness. We can overlook differences when we remember we are all connected to the same Source, we all come from the same Source, and it's to the same Source we return.

Chapter 15

Practicalities

'This may only be a dream of mine, but I think it can be made real.'[1]

Ella Baker

So far, this vision of 2197 may seem rather metaphysical and abstract, even utopian and idealistic. You might be wondering how we actually put these lofty principles into practice. High-minded ideas are little use if they can't be grounded into the day-to-day practicalities of real life, but the principles of coevolution, courage, connection and compassion are eminently actionable, and capable of informing all aspects of human society.

Let's revisit the societal structures from the chapter on Power to see how these concepts might be embodied in the new civilisation. At the end of each section I summarise the key values that underpin the vision.

These are suggestions rather than stipulations; you may have better ideas or complementary ideas – in which case, I invite you to share them and start bringing them into form. We're creating this future together.

Environment

Imagine a time when the word *environment* has become irredeemably old-fashioned. It has fallen out of common usage because of its implication that the environment is something outside of and separate from humanity, which of course is a crazy notion. We eat, drink and breathe the environment. We live in it, and it lives in us – our gut flora inhabit us, just as we inhabit the Earth – except that we need our friendly bacteria more than Earth needs us.

For all practical and spiritual purposes, we are inseparable from the environment. We can't draw a line between where we end and the environment begins. When we understand that, we are as invested in caring for the natural world as we are in caring for our own bodies.

Indeed, we need to care for the natural world a great deal better than we currently take care of our bodies. It's amazing we aren't even more diseased than we are, given the junk food, air pollution, substance abuse, pharmaceuticals, toxic cosmetic and hygiene products, pesticides and high frequency electromagnetic pulses that we expose ourselves to.

Environmentalism used to mean caring about one's environment, and used to mostly focus on conservation of natural resources – air, oceans, forests and species. It evolved into *sustainability*, which was primarily an anthropocentric perspective, seeking to ensure that future humans would be able to enjoy a standard of living on a par with current humans. Then there came *regeneration*, positive and proactive healing of the Earth, restoring the land, air and sea to pristine condition.

In the new civilisation we enter the phase of *coevolution*. All previous phases had been about humans doing something *to* nature – conserving it, sustaining it, regenerating it. Subject-verb-object. Now, with our acknowledgement of the deep interconnectedness of all things within Gaia, the emphasis shifts to working intimately *as part of nature* to coevolve. Humans partner with the creatures, plants and even the rocks and mountains, in reverence for the beauty and power of nature.

In our imaginary world of 2197, all products are designed on closed-loop, cradle-to-cradle principles of repairability and recyclability. Most consumer and household goods are easy to repair, if not by the community that co-owns them, then by one of the countless repair shops that can fix just about anything. Items that are unsuitable for repair are fully recyclable, right down to the molecular level. Everything is broken down into its constituent molecules and reincarnated into new products. Nothing, absolutely nothing, is discarded, apart from compostables that fully biodegrade and nourish the soil.

We currently have this word, *disposable* – as in disposable gloves, masks, nappies, wipes, cameras and vapes. That's another word, along with *environment*, that becomes obsolete from this new mindset. Why would we want to *dispose* of something? And where would we dispose of it *to*? Even if we launched it into space, it would still exist somewhere, and once we see ourselves as profoundly linked to the entire cosmos, we wouldn't litter space any more than we would litter our own home. Plastic used to be seen as hygienic because it could be used once and thrown away. In the future it seems the height of *unhygienic* behaviour to chuck into landfill a single-use item that has an afterlife of 100 years or more.

Centuries from now, people will still be digging up our so-called disposables. They will be gradually working their way through the landfill sites, processing the contents through the molecular recycler to reclaim and reuse the plastics, minerals and metals that we so short-sightedly discarded literally as if there were no tomorrow.

We will certainly still be living with the consequences of global warming, no matter who or what caused it. Once the climate passes the tipping point into a new normal, it is likely to stay stuck there for a while. A Kali-type event might cause the collapse of the human population, or a similar collapse could be caused by the most massive own goal in human history: changing our climate to one that doesn't support life as we know it. But even if only a few humans survive, there's hope. The geological record suggests that around 70,000 BCE, the human population plummeted to between 5,000–10,000 individuals.[2] Yet here we still are.

Remember the IPAT equation: environmental Impact as the product of Population, Affluence and Technology? An event like Kali would cause all these variables to change dramatically.

Population would decrease as people die of starvation and disease. This would give humanity the opportunity to intentionally keep our numbers low and steady, to leave more of the Earth for non-human species while bringing us back within Earth's carrying capacity.

We could redefine affluence to mean wealth of emotional riches, rather than the meaningless acquisition of material stuff.

And Technology could also come to mean something very different, consisting not of nuts and bolts, or even circuit boards and microchips, but social, medical and agricultural technologies inspired by and in harmony with nature, along with free energy devices and innovative transportation methods.

These shifts in the IPAT variables would facilitate a fundamental change in the balance between humans and the rest of nature. Humans, our pets and livestock currently account for 96 per cent of all mammalian biomass.[3] In the future, we don't need as many domesticated cattle, pigs, sheep and goats because a healthier, primarily plant-based diet has become the norm, and there are far fewer of us. The ratio is reversed, with free animals, living as nature intended, comprising the majority, humans the minority. Herds of elephant, giraffe, zebra and rhinoceros roam the savannahs of Africa; wolves and polar bears ply the Arctic tundra; bison, beaver, lynx, ocelots and porcupines throng the North American plains. They have plenty of space and food, so they mostly don't bother the humans. It was only when they were under threat of starvation that they resorted to desperate measures like venturing into human settlements. After centuries of persecution, poaching and exploitation, they still carry the collective memory that it's wise to be wary of *Homo sapiens*, so they mostly keep their distance. But that wound is gradually healing, and life-affirming stories emerge of special encounters and enduring relationships between humans and our non-human kin.

There is lower demand for travel and transportation, because food and goods are produced close to where they're used, people live closer to their families and friends in close-knit communities, and there isn't anywhere near the same demand for travel-as-escapism. We can imagine a world of radical new transportation technologies – vacuum tube trains for long journeys, fleets of collectively owned autonomous flying vehicles for local travel, and drones for transporting goods. Maybe we've even figured out how to harness energetic frequencies for teleportation. The cumulative effect could be that roads have become obsolete, so wilderness is no longer criss-crossed by the web of roads and motorways that used to break up the roaming ranges of wild megafauna, trapping them in smaller and smaller areas and disrupting their natural migrations.

Urban habitations become more vertical, less sprawling than they used to be, using new high-strength construction materials made from mycelial fibres like fungus, or calcium carbonate like seashells. They have an organic lightness and elegance, and a proliferation of greenery on rooftop terraces and living walls, inspired by the futuristic solarpunk aesthetic. Rural communities are generally small, and designed to blend harmoniously with the landscape. Instead of pockets of nature surrounded by humanity, we now have pockets of humanity surrounded by nature. The world has become a more beautiful place.

Nature rapidly reclaims the towns and cities that are left deserted after Kali. Within forty years of the Chernobyl nuclear power plant disaster of 1986, animals and plants were colonising the empty cities of Chernobyl and Pripyat, thriving amidst the rubble.[4] Post-Kali, trees push through roofs, ivy demolishes walls, birds nest in old office buildings and nature's tiny recyclers – the worms, ants, fungus and rot – determinedly set about converting derelict human habitations to dust.

Rising oceans lay claim to coastal and riverside areas. Sea levels rise nearly 30 feet higher than they were in the early twenty-first century, so cities like Jakarta, Venice, Rotterdam, Miami, Guangzhou, New York, Virginia Beach, New Orleans, Lagos, Calcutta, Shanghai, Mumbai, Tianjin, Tokyo, Hong Kong, Ho Chi Minh City, Bangkok and London are at least partially underwater.[5] A few intrepid souls dive the underwater neighbourhoods, and the submerged cities become marine protected areas, havens for ocean life. The oceans themselves come within a whisker of complete ecological collapse – acidification wreaks havoc with the plankton that form the basis of the oceanic food chain – but Kali's intervention gives them the time they need to recover. Nature is incredibly resilient – life wants to live – and she heals herself when simply left alone.

Principles: connection, balance, reverence.

Economics

The *veil of ignorance* is a term coined in the twentieth century to mean that policies should be designed as if the policy-makers didn't know where they might find themselves in society – high or low, rich or poor. Under the veil of ignorance, the invitation was to devise policies that were fair to all, no matter where they were in society's pecking order. It was a nice theory, but essentially never happened in practice. Most policy-makers were relatively privileged, and tended to bias towards self-serving policies that favoured the rich while making the poor poorer.

In the new civilisation, we go one step further than the veil of ignorance, to deep interconnection. Economic policy is based on the principle that if any one of us is prevented by poverty from meeting our basic needs, it diminishes our collective capacity for conscious coevolution. If a rising tide raises all boats, poverty holds back the tide more effectively than Canute ever managed to.

The old economic model was founded on a principle of scarcity. The new economics is founded on the principle of abundance. If scarcity is the gap between what you want and what you have, abundance is the sense that you have more than you want – and in the new civilisation we have more, and we want less.

It's easier to feel abundant now that there's a healthy balance between humans and the rest of nature, with plenty of space and resources to go around. Excessive consumption is seen as seriously passé, while altruism and generosity are the new normal. After Kali, the remaining humans knew they needed to band together if they were going to survive, so it was in their interests to keep everybody alive by sharing resources. If anybody had more than they needed, they gave the rest away – *I store my spare grain in the belly of my neighbour*. Now that lifestyles are consciously simple, and most things can be grown, made or otherwise generated within our local communities, sufficiency is the new abundance.

Income distribution is now decoupled from what used to be called work, which was a strange, stressful and unfair way of allocating wealth. Everybody receives a universal basic income, called the Adequacy Allowance, from government. This enables people to spend time on enjoyable, creative projects that contribute to personal growth and happiness. Goods and services that sit uneasily alongside a profit motive are provided unconditionally and free of charge by the national Ministry of Essentials: education, healthcare, elder care, childcare, fire and ambulance services, police, water, energy, sanitation, refuse collection, recycling, public parks and museums. Housing and high-quality food are price-controlled to ensure nobody goes without. Historically, some people disliked the idea of public provision because it had been badly implemented – low quality or limited services, sometimes subject to oppressive conditions. This wasn't a failure of the idea, it was a failure of execution. When you can have access to the best possible services, free of charge and free of provisos, why wouldn't you?

But if you don't like what's on offer, you are of course free to create your own school, healthcare facility or whatever, using a government grant. Alternatively, you can request your local providers to provide whatever it is you wish for in addition to their existing services, and the community will determine whether there is enough support for the idea to make it worthwhile.

Robots do the jobs that nobody wants to do but that need to be done, although there are very few such jobs. The dirty jobs used to be mostly involved with disposing of society's waste substances – collecting rubbish, recycling toxic materials, maintaining sewers, and so on. Now there is effectively no waste, so those jobs have gone away.

You might wonder why people don't just laze around doing nothing if they don't have to work for money. That would be to underestimate the power of the evolutionary impulse. Throughout all of human history, people have wanted to discover, invent, build and grow. You can see this in children – they are agog with curiosity to explore the world, eager to create. It's part of what makes humans human. But we used to live in a system that told us we had nothing valuable to offer, or associated maturity with giving up on youthful dreams and buckling down to a proper grown-up job. It's small wonder many people became discouraged and gave up. The new society is designed around supporting people to follow their heart's desire and bring their gifts into the world.

Money doesn't occupy a central role in our society in the way that it used to. Chasing money for its own sake seems pointless, a distraction from the higher goal of learning and growing. Each community makes its own policy decisions on this, but most have a cap on wealth accumulation. The Adequacy Allowance is enough to live on quite comfortably, and they know that an unequal society makes *everybody* less happy, including the rich. The ceiling on income is 10xAdequacy, but it's mostly a theoretical limit. The social taboo on greed means that most people start giving away their surplus once they get to 2x or 3xAdequacy.[6] There will always be some who want to accumulate, but they are usually regarded with compassion rather than envy, as they're clearly trying to compensate for some deeper lack.

We have a variety of different currencies for different purposes – more of a polyculture than the current financial monoculture. As with monoculture farming, having just one currency creates systemic vulnerability, always booming or busting. If all our eggs are in one basket, be that an agricultural basket or an economic basket, if that basket falls and our eggs break, we've got a serious problem. So we have local and global currencies, currencies for necessities and currencies for luxuries, a currency that is used for the Adequacy Allowance and a different currency that is used for selling the fruits of our labours. Each currency can be tweaked so it creates the most desirable system of incentives for its specific role. It's hard for the people of the future to imagine how we used to manage with one-currency-fits-all, not just from the resiliency perspective, but also because these different uses for money have quite different purposes, and correspondingly different design requirements in terms of geographical coverage, fast or slow transaction rates, fungibility, and so on. This might sound terribly complicated, but digital wallets take care of the complexity, and the benefits of stability more than justify the new currency model.

Most importantly, money is no longer weaponised as a tool of power and domination by individuals, companies or governments. Interest-bearing loans are illegal and branded as usury, as they were in England until the thirteenth century.[7] With stable currencies and no inflation, there is no valid reason for charging interest. Rather than asking a bank for a loan, anybody needing to fund a new enterprise applies for a government grant, which is almost always approved, and includes unlimited consultation with skilled advisors. This halts the scarcity-generating game of musical chairs that used to cause such misery for individuals and nations alike.

There is, of course, more to economics than these few examples but, from a general perspective, the economic system is elegantly simple compared with that of the twenty-first century. Systems tend towards complexity, and by the time of Kali economics had become so complex that it was understood by very few people, if any. This meant that the so-called experts were accountable to nobody, because nobody understood what they were doing, often not even the experts themselves. In the future this deliberate simplicity frees up an enormous amount of headspace to focus on the important things in life. The economy has been put back in its place: it exists to serve humanity, rather than humanity being enslaved to the economy.

Principles: equality, abundance, adequacy.

Politics

Politics as such no longer exists, certainly not party politics. It was a massive waste of time, money and effort to have Party A in power for a while, only to be replaced by Party B, which would undo much of Party A's work, at least until Party A got back into power and undid Party B's work. Most of the Indies hadn't engaged in party politics for a long time before Kali, being disillusioned not just with political parties, but with the entire corrupt, corporatocratic system.

The world has now moved on from politics, with all its patriarchal baggage, to *governance*. The word governance comes from the Greek *kubernan*, 'to steer', embodying the idea of route-planning and navigating the ship of state, rather than commanding like a captain. To this end, the Indies created a stable, transparent, accountable central ministry responsible for providing public services as listed above. A bit more etymology: 'to minister' originally meant 'to act under the authority of another', meaning that a ministry's role was to defer to the authority of the people. Somewhere along the way, this got turned upside down (as so many things did) so that people were deferring to the authority of the ministry. The Indies set things the right way up again.

The new societal organisation places trust in the people, devolving as much power as practically possible to ordinary citizens (no longer known as *consumers*) to run their own communities. The building block of governance is the Community Council, a people's assembly based on indigenous council practices.

Everybody serves on the Council at various points during their lifetime, for a twelve-month term on a part-time basis. These governance sabbaticals are based on the Fibonacci sequence as a reminder of the mathematical perfection of nature: at age 8, 13, 21, 34, 55 and 89. With the new healthier food and lifestyles, a few even serve at age 144. Age 8 may seem young to have a role in governance, but the participation of the young ones is crucial. They bring an innocent and clear-eyed perspective to meetings – not to mention disconcertingly pertinent questions, often challenging the adults to see things a different way. The younger Councillors are assigned to an elder in a kind of apprenticeship role – the elders have valuable life experience and historical context, and over time they impart this knowledge to the youngsters so they're not fated to repeat the mistakes made by previous generations. It's a mutually respectful and beneficial exchange of perspectives that strengthens the interpersonal fabric of the whole society.

It's not compulsory to serve on the Council – it would be contrary to the principle of personal autonomy and freedom to force anybody to serve against their will – but most people are honoured to contribute to the governance of their community and willingly accept the invitation. Individuals of the relevant ages are selected by lottery, adjusted for gender, race and other characteristics to ensure genuine diversity of representation.

As well as the regular Councillors, every Council also has a designated Joker, and a representative for the Children. The Joker's job is to be contrary, flippant, argumentative, or just plain silly. It's very important to have a Joker in the pack to save the Council from lapsing into conformity and groupthink; it encourages lateral thinking and fresh perspectives. And a few belly laughs liven up Council meetings.

The role of the Children's representative is to remind the Council to consider the welfare of future generations, and to remember that their decisions will have far-reaching impacts. The representative doesn't have to be a child – in fact, the older they are, the more years they've had to witness how decisions play out in the long term. The position is typically occupied by an elder.

The pre-Kali method of governing was very left-brained and matter-based, full of policy documents, budgets and cost-benefit analyses, produced by a relatively homogeneous group of bureaucrats. Apart from being incredibly dull, which discouraged a lot of people from getting involved, it also skewed decisions towards things that can be measured, rather than things that matter. We ended up with one-dimensional decision-making processes that missed out much of what makes life worth living.

Fortunately, many of the Indies were members of indigenous tribes or intentional communities, so they already had strong skills in decision making and governance through collaborative, consent-based processes. Their practices entail a more holistic, transparent and inclusive approach. They use exercises in imagination and creativity, such as systemic constellations, to achieve a rounded picture of consequences both intended and unintended. In a constellation, Councillors take on various different roles in relation to the issue under discussion. As well as the Joker and the Children, these might include a representative for Water, who reflects to the Council how a policy will impact on water in all its forms – rivers, oceans, rainfall, domestic water supplies and so on. Other roles might be Trees, Soil, Birds, Earthworms, a neighbouring community and any other stakeholders that could be impacted by the new policy. This role-playing often reveals surprising and important perspectives that influence the final decision.

You might wonder how there are enough roles for everybody, because in the old order we were used to being governed by a *small number of people* representing a *large number of people*. In the new democracy, units of governance are significantly smaller – three times Dunbar's number, or 450 people. Robin Dunbar defined his number as 'the number of people you would not feel embarrassed about joining uninvited for a drink if you happened to bump into them in a bar'.[8] In other words, the maximum number of people a human can know well enough to remember who they are and how they are related to other people within the group. Three times this number is large enough to achieve a critical mass but small enough for everybody outside of the Council to feel they are being represented by someone they trust, and everybody inside the Council to feel personally accountable to the community.

There are still disagreements – humans are still human – but we've developed ways to disagree respectfully. When people respect the process, they may not agree with the outcome but they accept it because they can see that due process was followed.

There are some decisions that need to involve larger demographics for practical reasons, so decisions are filtered up the hierarchy to the appropriate level. Most decision making, though, is local, following Elinor Ostrom's guidelines on the management of the commons, especially the guideline that anybody affected by a policy should have a say in making that policy.[9]

Many physical communities are much larger than 450 souls – although there has been a drift from urban back to rural living, many people still live in towns and cities – but whether city or country dwellers, we still find it beneficial to subdivide ourselves into groups of about this size, comprising a number of complete households. Households tend to be larger than they were in the twenty-first century, consisting of extended families living in assorted dwellings on a shared property. Households had previously become very fragmented as people moved far from their hometowns to find paid work, which put a lot of pressure on the nuclear family, which in turn contributed to high divorce rates. Raising children in such a troubled society was hard, and it was more than many marriages could withstand. Believing that it takes a village, many of the Indies had already returned to a more communal way of living before Kali, finding it a generally better arrangement for child rearing, food production and reduction of environmental impact, so post-Kali that becomes the new normal.

Principles: localisation, representation, participation.

Education

Children should be treasured in any sane society. One day they're going to be running the show – and in post-Kali society they get involved in governance sooner rather than later – so why wouldn't you want them to grow up secure, happy, confident, healthy and with a clear sense of who they are and what they're here to do?

And if children are so crucial to the future thriving of a society, surely you would also want them to have the best possible educators during their formative years. If you're going to value the children, you also have to value the people charged with their care. Educators are among the most honoured roles in the future society.

The new civilisation doesn't see children as adults-in-training, workers-in-training or (worst of all) consumers-in-training, but as beautiful beings still close to the source of their life's river. As the collective consciousness of humanity rises, each generation is born a little more conscious than the one before. So the last thing we would want to do is teach children to be mini-versions of their parents and grandparents. That would be a tragic waste of their potential, not to mention a missed opportunity: adults don't want to teach them to be like us – we want them to teach us how to be more like them. So the role of education is not to force-feed knowledge, but to create a psychologically safe space for children while they learn, getting back closer to the original Latin meaning of *educare*, to lead out, rather than to put in.

Having said that, there are certain life skills that come with maturity and experience, as well as practical creativity skills, and these are important components of education. But the main objectives of school are to give children the self-confidence that will allow them to fully express their gifts, while also encouraging their natural curiosity about the world, exposing them to as many ideas and activities as possible to help them find the pursuits that make their heart sing and will enable them to make their best contribution to the future thriving of society.

It's easy to see the origins of the new education system in the Indie culture, particularly the contingent from intentional communities. Many of them had embraced the ideas of the nineteenth-century social reformer, Rudolf Steiner, as implemented in the Waldorf school system and various home-schooling methodologies. He believed that children are spiritual beings, born with creativity and inquisitiveness, and the main goal of education should be to nurture those qualities, not quash them. For the first seven years, children should be given to believe that the world is good (in order to build a sense of security), in the next seven years, that the world is beautiful (supporting appreciation and curiosity), and in the seven years after that, that the world is true (conducive to harmony, justice and truth), emerging ultimately from education as free, independent and creative beings.

Principles: nurturing, expressing, thriving.

Justice

Given this rosy picture I'm painting of twenty-second-century civilisation, you're probably expecting me to say there will be no crime, so no need for criminal justice – but that is unlikely to ever be so. Much crime is generated by poor systems design and artificial scarcity, but, even so, humanity is unlikely to ever become a species of angels, no matter how effectively we address the systemic issues.

Theft, robbery and fraud are motivated largely by desperation, envy or entitlement. If other people have money or stuff that you don't have and can't afford, you're going to be tempted to take by force what the system won't give you by right. Many crimes will be eradicated by everybody having what they need in order to live decent lives, and people not requiring material things to make them feel better in a world that appears to despise them. When your status comes not from what you own, but from who you are, you don't need to have a flashy car or cool clothes to create an impression.

Violence, rape and murder also drop. Crimes against the person often arose out of addictive behaviour like alcohol abuse, which in turn was a form of escape from a terrible life. Once basic needs are met, there is less dependency on drugs and alcohol to numb yourself to the pain, the feelings of failure and loss of dignity. When education builds self-worth, and good jobs provide a sense of being both valuable and valued, people no longer need to live lives of quiet (or violent) desperation.

But there are still crimes. Human beings have always, and will always, consist of the dark as well as the light. We need to know that we have the option to go to the dark side so we can exercise our free will and consciously choose the light, and there will always be some who choose the darkness.

So what happens when there's an encounter with someone who has chosen the dark, with tragic consequences? This may be hard to hear for anyone in the twenty-first century who has been a victim of crime, but an offence is a tragedy only at the personal, ego level. This is not in any way to diminish the impact of the pain and suffering, but to offer the bigger-picture perspective that *all* life experiences, both good and bad, can contribute to our soul's evolution if we choose to use them that way. A painful encounter with the forces of darkness can propel a person to a higher level of understanding. Heavily disguised as it may be, being subjected to a crime can be a gift to the soul. There can be post-traumatic stress, but as Stephen Joseph found, over time this can transmute into post-traumatic growth, in which the trauma leads to greater wisdom and insight.

This is not to say that the perpetrator should get off scot-free. It's important that the society has clear boundaries around what behaviours are conducive to the greater good and which are not. Consequences are crucial feedback loops.

The main purpose of criminal justice in the new civilisation is to support the personal evolution of the person who committed the crime. (Note that we don't call them criminals any more – it is the deed that is criminal, not the person.) Justice isn't about punishment, it's about the teachable moment, for both the perpetrator and the society.

First, what can the society learn? Going beyond the presumption of innocent until proven guilty, the presumption now is that society has failed in some way, rather than the individual, and society does its fair share of soul-searching to figure out how it could have better supported the individual who has gone astray. Are there systemic factors that led this person to behave in this way? What unfulfilled needs did they have? Is there a systemic change that could prevent this situation recurring in the future?

Second, what can the individual learn? What was their state of mind, their motivation? How could they have made a more positive choice? Do they regret what they did? What have they learned from this experience? Are they committed to not making the same mistake again? What lifestyle adjustments would support that commitment? Are they willing to make reparations to the wronged person or their next of kin?

This process looks a lot like therapy. Some perpetrators joke that they would prefer the old regime where they simply got banged up in prison. Imprisonment would have been easier, because this is the hardest work they have ever done. No corner of their psyche is left unexplored. The questions go deep, until the therapist, the perpetrator and the wronged person are all satisfied that they've got to the heart of the issue, that the wound has been surfaced and healed, and the perpetrator made whole again. It is rehabilitation at its deepest level, and can take almost as long as an old-fashioned prison sentence, but is much more effective – not just for the perpetrator, but for the society as a whole. A wound in an individual psyche creates a wound in the collective psyche, and the healing of the individual heals the whole.

Principles: collective responsibility, rehabilitation, reparation.

Technology

In an interconnected web of conscious agents, the level of consciousness of an agent influences the agents around it, and technology is a particularly powerful agent because it is so highly networked, touching so many lives. This is not to say the technology itself is conscious – before Kali knocked out everything electrical, even the most advanced artificial intelligence did not have a sense of its own self-ness – but technology embodies the consciousness, values and goals of its creators. Presenting tech as value-neutral is disingenuous. Before Kali, the Indies were already waking up to the potential toxicity of social media – with its algorithms of outrage and economics of distraction. Post-Kali, technological innovation is approached more mindfully.

There used to be a crazy urgency around innovation. Looking back at technological development in the early twenty-first century, the rate of progress was impressive, but also reckless. The long-term implications and consequences of new devices and applications were largely ignored. People struggled to keep up with the phenomenal rate of change as technology hurtled forward at breakneck pace, with little thought given to repercussions.

It was actually a mercy that Kali came when she did, as AI was right on the brink of eclipsing its human creators. It was getting truly dangerous. (Digital surveillance was getting out of hand too, but that was quite deliberate, a different kind of dangerous.) AI had got just about smart enough to do what it was designed to do, but was still too dumb to know when to stop. The philosopher's favourite – the paperclip problem, in which an AI tasked with making paperclips might cause catastrophe by commandeering ever-increasing resources into making paperclips, while also generating increasingly sophisticated ways to fend off attempts to stop it[10] – was in danger of coming true. Maybe humankind really had made AI in its own image, because it could be said that humans, too, were smart enough to do what they set out to do, but too dumb to know when to stop. There was a rumour that the tech companies realised AI had got away from them and somehow manufactured the solar flare as a desperate last-ditch attempt to stop it from taking over the world. However, the belief that we could control the Sun is as crazy and hubristic as the old-school belief that we could control the Earth.

This headlong, headstrong dash into tech-superiority was fuelled by the desire to get rich quick. One of the main defences of neoliberal capitalism was that it rewarded innovation, and it certainly did. There were no constraints on how excessively rich you could become, and no expectation that you would share your intellectual property. All the incentives were there to innovate, rush to market, make a killing and sell the business – ideally before any downsides came home to roost.

The Indies have a very different perspective. Open-source technologies were already getting traction before Kali, and after Kali it made even more sense to share knowledge, in keeping with the general sense of community, equality, gifting and interdependence. We would no more hoard intellectual property than we would hoard food, resources or money. It benefits everybody when we co-create and build on each other's ideas.

Because we're not in a race to cash in and cash out, we can afford to take a more leisurely, cautious approach to technology. We want to take the time to really think through what we're bringing into the world, and whether it will serve the greatest good in the longer term as well as the short term. After all, if we bring a bad idea into being, it might well be around the next time we reincarnate, and we'll reap our own whirlwind. There's little incentive to rush, and every incentive to observe the precautionary principle.

We're still not perfect at foreseeing unforeseen consequences. Humans never have been, and likely never will be, even with intensive modelling and carefully designed pilot schemes. So every new invention has to pass the so-called Genie Test – we don't let it out of the bottle unless we have a proven way to put it back in. This might sound stifling to innovation, but it isn't. Innovation has slowed to a more sustainable pace, but that is exactly the point. We would rather innovate slowly, and innovate better.

Paradoxically, this seems to generate more significant advances with each new release cycle. When profit and market share were the goals, manufacturers kept the money rolling in by designing for obsolescence and releasing new versions with tiny incremental improvements over the previous version for the dedicated consumers who just had to have the latest model. Even though (or maybe because) the technical advances were so small, an enormous amount of money, time and energy was spent on marketing the new product. They had to get people excited about it somehow.

Now all that creative energy can be directed towards actually making the product better, safer, more elegant in its simplicity and suitability. There might be longer gaps between new versions coming to market, but they are worth waiting for and they don't require sophisticated marketing campaigns to *persuade* people that they are better because they simply *are* better.

Principles: caution, quality, reversibility.

Media

The stories that a civilisation tells itself about itself are crucially important, and this is the role of the news media. You could even say that the news media form a society's self-image, because journalists get to choose from the vast array of possible stories which ones to reflect back to a society. Media becomes the mirror. When a society looks in the mirror, what does it see? This becomes its reality.

Pre-Kali, news media portrayed a horrifying picture of murder, war, rape, violence, crashes, corruption, health scares and over-privileged people generally screwing up their lives and screwing up the planet. Was life really like that? Or is it just that the media chose to reflect those stories? But why would they want to make a society feel so appalled by itself?

The indigenous people who survived Kali brought a healthier tradition of cultural storytelling – indeed, their stories helped the Indies to get through the tough times immediately post-Flare. Their stories focus on courage, resourcefulness, redemption, surviving hardship, overcoming obstacles, and the power of love. That's not to say they ignore the dark side of human nature, but they see a story as incomplete until it reaches a point where something edifying comes out of the suffering.

Truth holds high value in post-Kali society, and truth requires nuance and respects complexity. Over-simplistic reporting, particularly the kind that glibly identifies heroes and villains, is rarely accurate. All news reporting is necessarily subjective, filtered through the personal lens of the reporter, but in the new civilisation self-awareness helps reporters be aware of their biases, and not to express opinion as fact. Journalism goes deep and true, rather than broad and sensationalised.

This kind of journalism obviously takes more research and reflection than jumping on the latest news bandwagon or lazily regurgitating corporate or political press releases, but post-Kali we're more careful about what we put into our minds, so we'd rather wait a while and read an article that is well-researched, truthful, nuanced, instructive and constructive than something that's deliberately crafted to appeal to our baser instincts. We're not interested in facile scapegoating, stereotyping, cheap shock or outrage.

The independence and integrity of the media is now regarded as so essential to a well-informed citizenry and effective democracy that it falls under the remit of the Ministry of Essentials, which provides unconditional funding, empowering journalists to do their best investigative work. Journalism is legally protected from government interference, and journalists are positively encouraged to hold government accountable – although with the new, more distributed, less corruptible governance model, this is rarely necessary.

Advertising as such doesn't exist any more; it's gone back to the pre-Bernays era of informative announcements of new products, rather than attempts to sell us glossily packaged snake oil on the pretence that it will make us healthier, wealthier or sexier. As for market share, readers are more discerning. They prefer quality to speed. As with technology, the pace has been taken down several notches, giving everybody time to breathe.

If that all sounds rather boring, it isn't. There's no shortage of drama and contrast in the world. That's the point of being human – to create opportunities to grow and evolve, which depends on change and dynamism. Because we're always pushing ourselves to try new experiences, they go wrong as often as they go right, and it's important that these stories get told so we can be inspired by and learn from others.

Comedy and satire thrive more than ever. Laughter feels good, it *is* good. Rather than *if it bleeds, it leads*, the news editors' philosophy is now more like *if you laugh, it leads*. News outlets renowned for their sharp wit attract the most readers. Humour, heroism and honesty are the criteria by which people choose their news sources.

Principles: truth, independence, integrity.

Science and Medicine

As you'd expect by now, having read about the slower rate of change in other areas, post-Kali we're much more cautious with medical and scientific innovations. Just because we can do something doesn't mean we should, especially when it comes to the human body, which we now know to be far more than just a sack of meat and bones.

The multidimensional nature of the body was well known to the ancient Egyptians, Chinese and Indians, among others, but got forgotten in the flurry of excitement about the new material sciences that emerged during the Enlightenment. The shift from a spiritual to a materialist worldview kicked off the most egregious and widespread case of confirmation bias in human history; when you assume that matter is all that matters, you create measuring instruments that measure only matter, which in turn justifies your assertion that all non-material phenomena are irrelevant because 'you can't measure it, so therefore it doesn't exist'!

This was a false dichotomy between science and spirituality, failing to recognise that they are not clashing, but complementary. Both are valid descriptions of reality, but different dimensions of reality. Physical ailments can originate in the emotional, mental or spiritual bodies, so trying to cure them in the physical body alone fails to get to the root of the problem and the disease returns. What in the twenty-first century is called *modern medicine* is really *material medicine*, and in the post-Kali world there is a much more holistic approach. The old medical practices now look as barbaric as leeches, blood-letting and trepanning once did – slicing and cutting without any regard for the subtle energy meridians that could have been used to cure the ailment, but instead get damaged by the bumbling, blasé butchery.

The approach to mental health is equally transformed. The early twenty-first century saw unprecedented levels of anxiety, mental illness and depression, often leading to suicide. Because we had this view of ourselves as disconnected beings, we saw this pain as specific to the person who was suffering from it, so we tried to cure it at the individual level by administering pharmaceuticals and treatments. This rarely worked, and in making the sufferer believe that they were defective in some way almost certainly made their suffering worse. They were, in fact, sensitive souls picking up on the trauma of preceding generations who had experienced countless wars, persecutions, violations, enslavements and other forms of brutality, through which both brutalisers and brutalised suffered psychic wounding. Although drugs might mask the symptoms, healing could never fully work at the individual level; it has to take place at the level of the whole, by venturing into the dark places of the collective psyche to transmute and cleanse our shared trauma. This work does, by necessity, have to be undertaken by individuals, but now with the understanding that they are not faulty or broken, but are doing this work on behalf of the collective.

Equally radical is the shift from medicine as disease management to medicine as wellness management. Lifestyles become healthier – we move our bodies more and consume higher quality food plucked fresh from a nearby community garden. We also don't make a habit of spraying our food with poison in the form of pesticides – why on earth did that ever seem like a good idea? We will still be cleaning up a lot of the toxins inherited from the twentieth and twenty-first centuries – there's a good reason why Persistent Organic Pollutants are known as *forever chemicals*. We obviously don't do that any more, not just because it's bad for us, but because it's also bad for plants and animals – and not just the ones the pesticides were designed to eradicate, but organisms throughout the entire ecosystem. We find ways to collaborate with and accommodate the beings formerly regarded as pests, rather than aiming for wholesale slaughter.

Human death, when it comes, tends to come swiftly. A soul decides it has evolved enough for one lifetime, and it's time to step out of the virtual reality game and return to the disincarnate state for some rest, recuperation and integration. Most people choose to live a long time, because there is so much that is beautiful in this strange mortal existence, and because we're curious to experience as much as we can before we die. But we certainly don't fear death. We see it as just moving from one dimension into another. We've come a long way from the twenty-first-century desperation to cheat the Grim Reaper. It's amusing that people were so busy trying to become immortal that they overlooked the fact they already were.

In our fictional post-Kali world, the cryogenically preserved bodies don't survive the Flare. Like everything else that depended on old-style electricity, the cryogenic capsules fail and the corpses thaw and rot. Physical death gets everybody in the end.

Principles: holistic, healing, wellness.

Religion

By now it will come as no surprise to you that organised religion goes the way of all the other power structures of the pre-Kali world. As with politics and economics, the time was long overdue for people to take power back into their own hands. We have never needed intermediaries to interpret and intervene for the divine. We've always had direct access whenever and wherever we wish.

The essence underlying religion has always been good and true. The perennial philosophy revolves around the divine, harmony, goodness and love. Religion itself has given rise to some of the most inspiring music, poetry, art and architecture on the planet, and has been a great comfort to many people, especially in times of need. But like any other large, manmade system, it became tainted by human ambition and lust for power. In the new paradigm, we know we can access the divine directly, know that we *are* divine, so we don't need the hierarchies that so often ended up being obstructions. Religious practice has become genuinely sacred, without the politics, paraphernalia and perversions.

The old religious buildings are preserved for their beauty, and because many of them incorporate timeless symbols of the divine – mandalas, rose windows, sacred geometry and divine proportions – and they are imbued with an atmosphere of peace and calm generated by centuries of prayer and devotional intent. Even though many of the old rituals seem archaic now, the churches, mosques, temples, cathedrals, synagogues, chapels, tabernacles and shrines are perfect sacred spaces for people to gather – and the acoustics are superb. We just had to do some deep spiritual cleansing to purge the residual vibrations of exceptionalism, judgement, prejudice, persecution and patriarchy.

Principles: direct relationship, sacredness, communion.

Chapter 16

Present

> 'Challenge and adversity are meant to help you know who you are. Storms hit your weakness, but unlock your true strength.'[1]
>
> Roy T. Bennett

Welcome back to the twenty-first century, with all its messes and imperfections, yet still a wonderful world, brimming with potential. Where do we go from here? How do we actualise the potential of the future?

I have no idea whether a Kali-like solar flare will happen, nor am I suggesting that I want it to. But history suggests that a major solar disruption will happen at some point, and there are many other natural phenomena that could equally spoil the party: pole shifts, asteroids, supervolcanoes, sudden climate changes (anthropogenic or otherwise) and more.

As I write this, the world is in the aftermath of Covid-19, not yet knowing if this is the beginning of the end, or only the end of the beginning of global pandemics. This has been a kind of disruption, but hasn't significantly altered the power structures. If anything, it has reinforced and exacerbated the existing ones; the rich have got richer, the powerful more powerful.

I must be careful what I wish for, but I believe that humanity *needs* to be disrupted. Paul Gilding writes:

> 'It takes a good crisis to get us going. When we feel fear, and we fear loss, we are capable of quite extraordinary things.'[2]

None of us knows the future, and we would be wise to mistrust anybody who claims that they do – but my guess is that the disruption is coming sooner rather than later. I feel this expectation with a mix of trepidation and excitement. We may well be the generation that has to go through the storm of change, and if not us, then our near descendants.

Of course, I would prefer to do this the easy way rather than the hard way, but increasingly it looks like that won't be an option. Ideally, we would wake up from our sleepwalk towards the abyss and change direction. Everything is possible, but several thousand years of human history would seem to suggest that as a species, we're much better at shutting the stable door just after the horse has bolted.

Who knows whether it will be an act of God that brings the crisis, or our own cupidity and stupidity? Or you could argue that the two are not mutually exclusive. You could say that, on a soul level, we know that we need adversity in order to grow, so we are subconsciously precipitating a crisis. For great growth, we need great adversity. The Kali Flare is my metaphor for this adversity, whatever shape it may take.

As well as death, time and transformation, the goddess Kali is also associated with liberation and spiritual release, the great and loving primordial Mother Goddess. She giveth and she taketh away. A loving mother wants the best for her children, which sometimes requires tough love if they are to mature into wise and self-reliant adults.

I doubt that humanity will step into the New Paradigm without the irreparable breakdown of the old. It would be like skipping straight from summer into spring, and nature doesn't work that way. There *has* to be autumn and winter – the fall, the dark time, the composting, the resting and regrouping – before new life can emerge. We're not good at accepting this fundamental law. We want eternal day, everlasting summer and a never-ending upwards trajectory. Western culture does not respect the cycles of life, doesn't want the darkness, the death, the decomposition before the return of light and warmth in the spring. But there is no escaping it; we can run but we can't hide.

I could, of course, be completely wrong. Maybe we'll wise up and rise up and do the right thing. There may yet be protests and pitchforks on a scale the world has never seen, and the uber rich may be forced to relinquish their power – but if that is how it goes, we are perpetuating the same old dynamic of might making right, only the power has shifted from the few to the many. It's still a pattern of domination.

So what to do? Do we sit around waiting for something to happen, whether it's mutiny by the masses or an existential wake-up call from Mother Nature? No, we can take active steps to make ready. In *Emergent Strategy*, adrienne maree brown writes:

> 'We will face social and political storms we could not even imagine. The question becomes not just how do we survive them, but how do we prepare so when we do suddenly find ourselves in the midst of an unexpected onslaught, we can capture the potential, the possibilities inherent in the chaos, and ride it like dawn skimming the horizon?'[3]

I agree with her that chaos is not something to be feared, but rather, embraced as a force for change. Einstein said that the most important decision we make is whether we believe we live in a friendly or hostile universe.[4] I'm opting for friendly, believing that everything that happens, no matter how tumultuous, will ultimately be for the greater good – although 'ultimately' could take a while.

Rather than fearing what I don't want, I'm focusing on what I do want. Like a surfer training to catch the perfect break, I'm building the skills, sense and sensitivity to ride this wave of change. I believe my soul has chosen to be part of this great transition. And I believe you have too, along with the 8 billion other souls alive at this time.

You might counter that it is precisely *because* there are so many of us that we've precipitated this crisis, by overpopulating and overburdening the Earth. We can choose to see it like that, and indulge in blame, shame and self-flagellation, but that doesn't help. Quite the opposite – that attitude will make us depressed, lethargic, apathetic and likely to drag down the spirits of those around us.

Many seemingly bad things have happened in my lifetime, on my watch – besides over-population, also extinction, pollution, oppression, inequality and so on – but what if there really is a purpose to all this? What if, in the grander scheme of things, all this had to happen in order for humanity to learn how *not* to live, so we would know how *to* live?

Just about anything of value that I've learned in my life, I've learned by doing it the wrong way first. And maybe that applies to us collectively too. We can choose to see this as a global rite of passage into our next level of maturity and, like most indigenous rites of passage (not to mention rowing across oceans), it *has to be* painful and scary in order for us to discover what we're capable of – to discover who and what we really are, and can become.

What if this isn't actually a disaster, but in fact the greatest evolutionary opportunity in all of human history? Barbara Marx Hubbard tells us that crises precede transformation. This perspective invites us to be heroes, rather than victims.

Preparing Without Prepping

There are important things we can do to make ready. These preparations serve a threefold purpose: they help us gain the skills we need to survive a crisis; they mitigate the impact of the crisis; and they sow the seeds of the future civilisation.

If you're recoiling at this, thinking it sounds prepper-like, I'm not asking you to stand on a street corner wearing a sandwich board proclaiming that the end is nigh, or investing in a lifetime supply of canned beans. Look at it this way: preparing for an uncertain future looks much the same as getting serious about reducing your carbon footprint and usage of virgin materials, or scaling back your reliance on broken and exploitative systems. Either way, it's a win. And it also feels good to know you can take care of yourself. Until the last 100 years or so, these skills were part of everyday life. It's only in the twentieth century that we in the developed world started relying almost entirely on buying stuff that other people have grown or made. People from the nineteenth century would laugh at how helpless most of us are.

Know Your Garden

To defer to the Hopi Elders' sage advice again:

> 'You have been telling people that this is the Eleventh Hour, now you must go back and tell the people that this is the Hour. And there are things to be considered …
>
> Where are you living?
> What are you doing?
> What are your relationships?
> Are you in right relation?

Where is your water?

Know your garden.
It is time to speak your truth.
Create your community.
Be good to each other.
And do not look outside yourself for your leader.'[5]

This statement offers valuable and practical advice. To me, this means creating a community of kindred spirits and working together to become as self-sufficient as possible. This doesn't have to mean communal living, although that's one option. It can equally be a band of like-minded neighbours from a village, town or city. Self-reliance entails a lot of work, too much for one person to do on their own, which is why you need to enlist a group of people you trust.

Assume that you won't be able to rely on supply chains of any sort – that goes for water, food, energy, medicine and material goods. Do whatever you can as a group to supply your own needs – grow fruit and vegetables, and learn how to preserve food without refrigeration by bottling, pickling, canning, fermenting, drying, curing or brining. Keep chickens if you want eggs, or goats, sheep or cows if you want milk, cheese and butter. Have equipment for cooking that doesn't depend on electricity or gas. Find a source of potable water that doesn't depend on an electric pump. Learn to forage. Study natural remedies using plants. Learn how to make and mend things like clothes, furniture and bicycles.

There are many beneficial side effects besides greater self-reliance. Studies show that working the soil releases microbes that trigger the release of serotonin, one of the happiness hormones. Gardening also improves cognitive function, life satisfaction and a sense of wellbeing.[6] Working alongside friends and neighbours towards a shared purpose builds local community, and one day you might need their help. The Amish have known this for a long time. Most of us in the West have got so used to thinking of ourselves as isolated beings, we forget that interdependence and interbeing aren't just spiritual niceties – they're practical necessities. Work with others to design a lifestyle based on principles of resilience and simplicity.

But, you might be thinking, not everybody can do this. There just isn't enough space in countries like Britain for everybody to be living off the land, and not everybody can afford to buy several acres. Is it right for *some* people to live this way if not *everybody* can live this way? It feels like looking after Number One, and the rest be damned.

This is a valid concern, but your suffering won't make anybody else suffer any less. It's increasingly obvious that human civilisation can't and won't carry on indefinitely as it is. If you and your community can take care of yourselves, you can take care of others. If you can't take care of yourselves, you have nothing to share with anyone else. Don't keep it a secret what you're doing – give others the chance to see, and to join you or follow your example. It's up to them whether they choose to do so.

Worst case scenario: you've spent all this time and effort, but the apocalypse never happens so you feel like an idiot, and I'll be right there alongside you with egg on my face. But from a different perspective, you'll have created a close-knit community, learned to tread more lightly on the Earth, and discovered a simpler, more fulfilling, joyous way of life. Could be worse.

If you're thinking you don't have time for all this because you spend most of your waking hours working for a big corporation, I get it. It is, of course, up to you to decide what's most important to you. If you really love working for a big company and you're comfortable with the ethics of all their practices, well and good. Do what brings you joy and will make you proud of a life well lived.

But if you've been thinking that you'd like a change of career, or to adjust your work–life balance, pay attention to that voice. It might not be a bad idea to accelerate the process. Spending more time with your family, or with your hands in the soil, or engaged in creativity, is likely to make you happier and healthier. What you sacrifice in salary you're likely to make up for in quality of life. When I quit my corporate job, I found that I'd actually been paying a great deal of money for the dubious privilege of doing the work I did – commuting, buying professional clothes, paying the mortgage on a house in London, treating myself to nice things and expensive holidays to cheer myself up after hard weeks and months at work. When I started doing work I loved, many of those costs were no longer necessary.

And while we're on the subject, you may also want to start changing your consumer behaviour, weaning yourself off the big manufacturers, energy companies and banks, and switching over to more local and ethical suppliers. As with simplifying your life and supplying your own needs, these changes will help ease the transition to a new way of living.

A New Model that Makes the Existing Model Obsolete

Buckminster Fuller said, 'You never change things by fighting the existing reality. To change something, build a new model that makes the existing model obsolete.'[7]

We need to stop trying to fix the old systems that are no longer fit for purpose. Withdraw your support from what is failing or broken. Invest it instead in creating systems for the new civilisation – new ways of living, new currencies, new belief systems ... Basically, a new narrative about what it means to be human, a narrative based on coevolution, connection, collaboration and compassion – compassion for yourself, compassion for all living beings, compassion for the Earth. *This* is the new model that makes the existing one obsolete. From this new narrative, everything else flows. Hold it as your truth, *inhabit* it, let it suffuse every aspect of your life and work. Make Donella Meadows proud: *this* is how we transcend paradigms.

I want to emphasise that a narrative isn't just about words – in fact, it's really not about words at all. Words are just vehicles for a *feeling*, for an *energy*. Words are a way that humans communicate feelings, but the words are just pointers, hieroglyphs, icons, representations in our human user interface. They're not the same thing as the emotional energies themselves, and it's the emotional energies that have the *real* power to create.

When I think about the obituary exercise that changed my life, I can't even remember now what words I wrote. But I can remember the feeling – of excitement, possibility, potential, freedom, fulfilment, lightness and a healthy kind of pride. I felt connected, plugged in, illuminated. I believe it was that energetic connection that created the shift, not the words on the page. In those moments when I was writing and feeling, I set wheels in motion in the metaphysical dimension. It took a while for the results to manifest materially, but I had written from my heart, I had set a conscious intention, I had felt the energy as if my fantasy were already true, and from then on it was just a matter of time. There were a lot of moments of doubt and fear, when I felt like I was having to summon up enormous reserves of courage. But, looking back now, I can see a kind of inevitability, an irresistible and inescapable momentum for change. Even when I couldn't see how I was going to get through seemingly impossible and impassable situations, I somehow always knew that I would.

I also learned that I didn't need enough courage to get to the other side; I just needed enough courage to pass the point of no return. After that, I'd figure it out as I went along.

The Bifocal Approach

There's a saying that we should see the world as it is, not as we wish it to be, but I believe we need to be able to hold both at the same time – both the world as it is *and* the world as we wish it to be; the truth of the now, but also the vision of the possible.

I call this the bifocal approach – through the top half of the spectacle lenses, I keep an eye on the horizon, and through the bottom half, I focus on what I need to do right this minute to get me a little closer to my destination. In other words, we hold the awareness of the world we want to create, but at the same time, we pay attention to the next step that we need to take, here and now, to get closer to that future potential reality.

Change may sound scary, but when I was afraid on the ocean, I would remind myself of the new life I envisioned for myself, and that helped steady me. I knew I wanted to look back and be proud of how I had shown up, consistently and stoically, no matter what storms were raging around me. This helped me keep the faith that everything was working out exactly as it should.

I discovered I was capable of far more than I'd thought. Humans tend to shy away from change, but we're actually very good at it. We've always been changing – but we mostly don't notice it because it's gradual, or it's a form of change that we regard as normal, even welcome. If you're a parent, think about having your first child – that's just about the biggest change that anybody can go through in a lifetime, and yet people all over the world do it, every day. In an instant you go from being a relatively independent and free person to having a squawling, red-faced, unpredictable and inarticulate little tyrant expecting you to meet their every need at all hours of the day and night, without even having the courtesy to tell you what they want. Yet most parents adapt to this new tyranny quite happily. All through human history, people have adapted to new circumstances which seem strange at first, but soon come to seem normal.

Just don't look too far ahead. It's easy to let our imaginations run away with us if we start to project too far into the future. Much better to take it as it comes, crossing each bridge as we get to it – or building a bridge where there is none – while keeping our eye on the prize.

The astrologer, Pam Gregory, tells a story from a time when she was learning martial arts and was set the task of chopping a thick builder's plank in two with her bare hand. Her teacher instructed her not to focus on the plank, which would hurt her hand, but to put all her attention on the spot 6 inches below it. Pam succeeded in karate-chopping her way painlessly through several planks. She felt invincible, and we can be invincible too – it's all a matter of focusing on where we want to end up, not on the obstacles, reefs and rocks that could shipwreck us on the way there.

We are resourceful and adaptable, and capable of more than we know. If it is figure-outable, I have no doubt we will figure it out.

Overcoming Fear of Death

The greatest fear for most people is dying, which is understandable. There are many ways to die that don't look much fun. And I can understand deathbed regrets. If we haven't lived life to the full, many dreams unfulfilled, we may fear dying with so much left undone.

I can also understand fear of hell. If you've been raised in a flavour of Christianity that preaches hellfire and damnation, and you've got more than a few doubts about the choices you've made, then being slow-roasted on pitchforks is an unappealing prospect.

But precisely *none* of the near-death or reincarnation experiences I've read or heard about bear out the existence of a hell realm. They say that there may be some purging and healing to do, but it's not forever, and not so bad. After all, if the purpose of the universe is to evolve, what would be the point of sending us to eternal damnation? That would be like throwing us into the evolutionary bin. Much better if we're guided through a learning process, then returned to life to make a better job of it next time around. At worst, it sounds something like school detention. Maybe hell, like detention, was invented as a threat to give a small number of people control over a large number of people, with clergy instead of schoolteachers.

If you don't believe in hell, but believe that when we die we simply cease to be, then what is there to be afraid of? There won't be a you to be aware of your own non-existence. As Mark Twain said, 'I had been dead for billions and billions of years before I was born, and had not suffered the slightest inconvenience from it.'[8]

Practice Letting the Heart Lead the Head

We may believe the brain is the most important organ in our body – but we have to take note of which organ is telling us that. I find it quite bizarre that we're aware of all the tricks the mind plays on us – the cognitive biases, fallacies and errors – and yet we still let it run the show. It's like putting a madman in charge of the world, despite being fully aware of his madness. Why would we do that? I'm not saying the rational mind is insane, although without the moderating effect of intuition or the cosmically connected right hemisphere, neither is it entirely sane, especially when it lives in a capitalist culture that exerts a strong selective pressure for sociopathy.

The left hemisphere, with its spotlight beam of attention, tends to focus too much on some things, not enough on others. When we perceive with the right hemisphere, or the heart, we bring the broader lantern beam. Developing this capacity helps us see the bigger picture, access more information as Michael Singer did when writing his doctoral paper, see the previously unforeseen consequences.

When we deliberately drop out of the mind and into the heart, we feel the peace that comes – the stillness, silence and spaciousness. I call it the Sacred Pause, a moment to remind myself of my true nature and remember that reality is an illusion. Everything goes quiet. I become very present. It's in this Sacred Pause that I discover I know things I didn't know I knew, can do things I didn't know I could do. For a time I escape the relentless hamster wheel of my mind and create space for whatever wants to come. I remember that life will give me what I need, rather than what I want. I see with the soul, not the ego; I witness with compassion, not judgement. I let the mind be the helmsman, but make the heart the captain.

There may be those who think my philosophy is akin to Pascal's wager,[9] or a desperate last drag on the hopium[10] pipe. Possibly they are right. I'm not suggesting we give up all efforts at climate change mitigation, or poverty alleviation, or any of the other worthwhile work going on in the world. I am suggesting that, *in addition to* that good work that keeps our left hemispheres feeling busy and useful, we also engage the heart and the right hemisphere to see the bigger picture, and engage on that level as well. We are a fractal of the whole, so by bringing ourselves into balance we bring balance to the collective. By healing ourselves we contribute to the healing of humanity's disconnection from each other and from nature. By accepting light and compassion into our own hearts, we bring it into the world.

Above all else, and without bypassing present suffering, it's essential that we hold a positive vision of how the world could be and believe that it can come to pass. Your vision doesn't have to be the same as mine – feel free to create your own – but whatever better future you envisage, hold that possibility in your heart with burning desire. Remember that everything that comes into physical being first existed as potential. We can bring a more beautiful world out of potentiality and into actuality through the power of lovingly held intention, love in the Buddhist sense of wishing all sentient things to be happy and well.

When we ask our soul about its deepest longing, and it presents us with a vision of what could be, it's easy to dismiss the vision as impossible, idealistic and utopian. We look at where we are now, and then at where we dream of being, and as we take the measure of the distance between them the gap seems too wide, our skills too inadequate, our will too weak. But once we've glimpsed that vision, nothing will ever be the same again. We can't un-see what we've seen, we can't take the writing off the wall, or put the genie back in the bottle. *The vision wants to come true*. When we envision a world of love, beauty, justice and peace, we have already started to achieve it by creating it first within our own selves.

For many years I've had an image in my mind's eye, an image that looks like a deep blue subterranean pool, into which I long to throw a big rock to send waves of change rippling out across the water. The pool is our collective consciousness, the dimension that connects us all.

As for the rock, that's you. That's me. We are all rocks with the potential to spread ripples. In fact, we're already doing so, but for the most part our ripples are unconscious and lack specific intention. When we recognise our power as catalysts for change, we become proactive, conscious actors influencing the course of human evolution. We show up and dance with life. We realise we are not just a drop in the ocean, but the entire ocean in a drop. And miracles just might happen.

Epilogue

'I searched for myself and found only God.
I searched for God and found only myself.'

Rumi

It was 2009, and I was rowing the second of three legs across 8,000 miles of Pacific Ocean. This stage would take me across the International Date Line and across the equator, from the west of the northern hemisphere into the east of the southern hemisphere.

I was between Hawai'i and the Republic of Kiribati, rowing through the doldrums, the equatorial region when the wind drops to nothing, the sun beats down and it feels hot enough to boil your brains. The day had been sweltering. The night was a little cooler, but it was still airless and stifling in the sleeping cabin. It was an unusually calm night, so I decided to lie out on the deck for a while, at least until the next rain squall came along. I dragged my sleeping bag out of the cabin and snuggled down between the runners of the rowing seat.

I gazed up at the stars. So far from light pollution, they were stunning. The Milky Way stretched diagonally across the sky. The longer I looked, the more stars I saw, until it seemed as if the entire sky was glittering with diamonds.

As I lay there on my little boat, I imagined the planets and moons orbiting all those stars, and for a precious moment I forgot everything. I forgot about my blisters, and my heat rash, and my aching shoulders. I forgot my nationality, and my gender, and the colour of my skin. I forgot who I was and where I was, and why. For a moment, I even forgot to be human, and simply allowed myself to be absorbed into the spellbinding beauty of the night sky. For a brief, incredible moment, I transcended my puny existence, feeling as tiny and insignificant as a mote of dust, but at the same time at one with the infinite majesty of the universe. I was everything, and everything was me. I was everywhere, and nowhere. I knew everything, and I knew nothing. I was eternal, and I was intensely present.

Then a squall blew in and I had to scurry back to the cabin, but the magic of that moment has stayed with me ever since. Compared with the vastness of the universe, our planet is tiny, our lives fleeting. In a sense, none of it matters.

And yet, all of it matters. This Earth is a miracle, the culmination of an elegant evolutionary process that has generated an incredible diversity of spectacular life forms, including you, including me.

To each of us, our lives matter. And to the Earth, even to the Universe, we matter too.

To Ponder

I have invited you to be a participant in this book. This isn't just a one-way communication from me to you. As you've been reading, hopefully you've had some emotional responses to my words – surprise, insight, inspiration, laughter – or disagreement, aversion, incredulity, indignation. I don't really mind either way – I just hope it's given you food for thought. Writing this book changed me. I hope that reading it changed you.

I'd like to leave you with some questions for further contemplation, with a journal, friend, in meditation, or in whatever way works for you.

What belief have you held that turned out not to be true? If that belief wasn't true, is it possible that you might have other beliefs that are also not true?

What stories do you believe about who you are and about how the world works? How are those stories working for you? Are there other stories that might work better?

Consider cognitive biases like confirmation bias, causality versus correlation, availability heuristics, hindsight bias, the bystander effect, the sunk cost fallacy and status quo bias. Where do you see those showing up in your life?

How do you feel about change? Excited or anxious? Resilient or vulnerable? If you feel fearful, what are the fears that arise? Write them down and examine them. Are they really so bad? If they still seem scary, try shifting from ego level to the level of an eternal soul – how do they seem now?

Do you notice yourself labelling experiences as *good* or *bad*? What if you saw them all as opportunities for evolution? Could that help make the bad times seem better, and ease the pain when good times come to an end?

Do you sometimes label people as ignorant, stupid or evil? Do you sometimes have an 'us and them' mentality? How would things change if you assumed that everybody is doing their best, given the resources at their disposal? Do you feel more compassionate? If you believed that everybody had been sent into your life with some message to deliver, or some lesson for you to learn, might you see your interactions more positively?

What do you think matters most in life? Fulfilment, happiness, contribution and relationships? Or hedonism, acquisition and amassing wealth? How happy do you feel in your life?

When you look at the world through the lens of domination and partnership, where do you see domination? Where do you see partnership? Where do you see 'othering' – of people, nature, or anything else that is to be 'conquered'?

Experiment with thinking with your heart, rather than your head. Regularly take a moment to *stop, drop (into your heart), feel*. Do you feel the stillness, silence and spaciousness there? Hang out there for a few moments, enjoying the peace and calm. If it feels good, consider making it a habit.

Imagine you're on your deathbed. What will make you proud of a life well lived? If you knew you were going to reincarnate onto this Earth at some point in the future, starting from the same level of consciousness at which you left it, how would that influence how you live your life?

You're welcome to share your responses with me via my website, at www.rozsavage.com/contact. I'd love to hear from you.

Further Reading

This is a list of books that have particularly influenced my thinking over the last two decades. Most have been referenced in the main text. All come highly recommended. I have arranged them by theme, although some necessarily defy categorisation. Where I have reviewed them on my blog, I include the link to my review.

Environment

Belton, Teresa. *Happier People, Healthier Planet: How Putting Wellbeing First Would Help Sustain Life on Earth*. Bristol: Silverwood Books, 2014. (www.rozsavage.com/narratives-neuroscience-and-no-money-favourite-books-of-2017/)

Hartmann, Thom. *The Last Hours of Ancient Sunlight: Waking up to Personal and Global Transformation*. London: Hodder Paperbacks, 2001.

Hawken, Paul. *Regeneration: Ending the Climate Crisis in One Generation*. London: Penguin, 2021.

Kimmerer, Robin Wall. *Braiding Sweetgrass: Indigenous Wisdom, Scientific Knowledge and the Teachings of Plants*. London: Penguin, 2020. (www.rozsavage.com/braiding-sweetgrass-kiss-the-ground-and-other-hints-of-hope/)

Quinn, Daniel. *Beyond Civilisation: Humanity's Next Great Adventure*. New York: Three Rivers Press, 2000.

Quinn, Daniel. *Ishmael: An Adventure of the Mind and Spirit.* New York: Bantam Books, 2009. (www.rozsavage.com/eating-ourselves-out-of-house-home-and-planet/)

Economics

Hickel, Jason. *The Divide: A Brief Guide to Global Inequality and its Solutions.* London: William Heinemann, 2017. (www.rozsavage.com/how-capitalism-is-failing/)

Jackson, Tim. *Prosperity Without Growth: Foundations for the Economy of Tomorrow.* London: Routledge, 2016. (www.rozsavage.com/tim-jackson-happiness-and-wellbeing-in-a-post-growth-world/)

Lietaer, Bernard, and Dunne, Jacqui. *Rethinking Money: How New Currencies Turn Scarcity into Prosperity.* San Francisco: Berrett-Koehler Publishers, 2013. (www.rozsavage.com/rethinking-money/)

Perkins, John. *Confessions of an Economic Hit Man.* San Francisco: Berrett-Koehler Publishers, 2004.

Peston, Robert. *How Do We Fix This Mess? The Economic Price of Having it All, and the Route to Lasting Prosperity.* London: Hodder Paperbacks, 2013.

Piketty, Thomas. *Capital in the Twenty-First Century.* Translated by Arthur Goldhammer. London: Belknap Press, 2017.

Politics

Watkins, Alan, and Iman Stratenus. *Crowdocracy: The End of Politics.* Romsey: Urbane Publications, 2016. (www.rozsavage.com/big-questions-democracy-for-the-future/)

Society

Frankl, Viktor. *Man's Search for Meaning.* London: Rider, 2004.

Gilman, Charlotte Perkins. *Herland.* London: The Women's Press Ltd, 1979. (Fiction)

Heffernan, Margaret. *Wilful Blindness: Why We Ignore the Obvious.* London: Simon & Schuster, 2019.

Norberg-Hodge, Helena. *Ancient Futures: Learning from Ladakh*. London: Rider, 2000. (www.rozsavage.com/favourite-books-of-2019/)

Storr, Will. *Selfie: How the West Became Self-Obsessed*. London: Picador, 2018. (www.rozsavage.com/whose-are-your-hoops/)

Spirituality

Chödrön, Pema. *When Things Fall Apart: Heart Advice for Difficult Times*. London: Element, 2005. (www.rozsavage.com/the-meaning-of-life-3/)

Dossey, Larry. *One Mind: How Our Individual Mind Is Part of a Greater Consciousness and Why It Matters*. London: Hay House, 2013.

Harvey, Andrew. *Hidden Journey: A Spiritual Awakening*. London: Watkins Publishing, 2011.

Hoff, Benjamin. *The Tao of Pooh*. London: Egmont, 2018.

Huxley, Aldous. *The Perennial Philosophy*. New York: Harper Perennial, 2009.

Redfield, James. *The Celestine Prophecy: An Adventure*. London: Bantam, 1994.

Rudd, Richard. *The Gene Keys: Embracing Your Higher Purpose*. New ed. London: Watkins Publishing, 2015.

Singer, Michael. *The Surrender Experiment: My Journey into Life's Perfection*. London: Yellow Kite, 2016.

Walsch, Neale Donald. *Conversations with God, Book 3: An Uncommon Dialogue*. London: Hodder & Stoughton, 1999.

Watts, Alan. *Tao: The Watercourse Way*. London: Souvenir Press, 2019.

Watts, Alan. *The Book: On the Taboo Against Knowing Who You Are*. London: Souvenir Press, 2009.

Woollacott, Marjorie Hines. *Infinite Awareness: The Awakening of a Scientific Mind*. London: Rowman & Littlefield, 2018.

Psychology and Neuroscience

Ferrucci, Piero. *What We May Be: Techniques for Psychological and Spiritual Growth Through Psychosynthesis*. New York: Jeremy P. Tarcher, 2009.

Gray, Dave. *Liminal Thinking: Create the Change You Want by Changing the Way You Think*. New York: Two Waves Books, 2016. (www.rozsavage.com/narratives-neuroscience-and-no-money-favourite-books-of-2017/)

Hari, Johann. *Stolen Focus: Why You Can't Pay Attention*. London: Bloomsbury, 2022. (www.rozsavage.com/stolen-focus/)

Hood, Bruce. *The Self Illusion: Why There is No 'You' Inside Your Head*. London: Constable, 2013.

Kahneman, Daniel. *Thinking, Fast and Slow*. London: Penguin, 2012.

Lotto, Beau. *Deviate: The Creative Power of Transforming Your Perception*. London: Weidenfeld & Nicolson, 2018. (www.rozsavage.com/narratives-neuroscience-and-no-money-favourite-books-of-2017/)

McGilchrist, Iain. *The Divided Brain and the Search for Meaning*. London: Yale University Press, 2012.

McGilchrist, Iain. *The Master and his Emissary: The Divided Brain and the Making of the Western World*. London: Yale University Press, 2009. (www.rozsavage.com/a-brain-of-two-halves/)

Taylor, Jill Bolte. *My Stroke of Insight*. London: Hodder Paperbacks, 2009. (www.rozsavage.com/a-brain-of-two-halves/)

Psychedelics

Bache, Christopher. *LSD and the Mind of the Universe: Diamonds from Heaven*. Rochester, VT: Park Street Press, 2020.

Pollan, Michael. *How to Change Your Mind: The New Science of Psychedelics*. London: Penguin, 2019. (www.rozsavage.com/beyond-the-doors-of-perception/)

Patriarchy

Beard, Mary. *Women and Power: A Manifesto*. London: Profile Books, 2018.

Blackie, Sharon. *If Women Rose Rooted: A Journey to Authenticity and Belonging*. Tewksbury: September Publishing, 2019. (www.rozsavage.com/if-women-and-men-rose-rooted/)

Criado-Perez, Caroline. *Invisible Women: Exposing Data Bias in a World Designed for Men*. London: Vintage, 2020. (www.rozsavage.com/favourite-books-of-2019/)

Eisler, Riane. *The Chalice and the Blade*. New York: HarperOne, 1988. (www.rozsavage.com/the-chalice-and-the-blade/)

Shlain, Leonard. *The Alphabet versus the Goddess: The Conflict Between Word and Image*. London: Penguin, 1999. (www.rozsavage.com/the-alphabet-versus-the-goddess/)

Chaos Theory

brown, adrienne maree. *Emergent Strategy: Shaping Change, Changing Worlds*. Edinburgh: AK Press, 2017.

Eoyang, Glenda H. *Coping With Chaos: Seven Simple Tools*. Cheyenne: Lagumo Corporation, 2009.

Waldrop, M. Mitchell. *Complexity: The Emerging Science at the Edge of Order and Chaos*. New York: Simon & Schuster, 1992.

Metaphysics

Currivan, Jude. *The Cosmic Hologram: In-formation at the Center of Creation*. Rochester, VT: Inner Traditions, 2017.

Gober, Mark. *An End to Upside Down Thinking: Dispelling the Myth That the Brain Produces Consciousness, and the Implications for Everyday Life*. Cardiff-by-the-Sea, CA: Waterside Press, 2018.

Hoffman, Donald. *The Case Against Reality: How Evolution Hid the Truth from Our Eyes*. London: Penguin, 2020. (www.rozsavage.com/just-how-real-is-reality-really/)

Kastrup, Bernardo. *Why Materialism is Baloney: How True Skeptics Know There Is No Death and Fathom Answers to Life, the Universe, and Everything*. Winchester: Iff Books, 2014.

Collapse

Diamond, Jared. *Collapse: How Societies Choose to Fail or Survive*. London: Penguin, 2011.

Gilding, Paul. *The Great Disruption: How the Climate Crisis Will Transform the Global Economy*. London: Bloomsbury Paperbacks, 2012.

Mails, Thomas. *The Hopi Survival Kit: The Prophecies, Instructions and Warnings Revealed by the Last Elders*. London: Penguin, 1997.

Ophuls, William. *Immoderate Greatness: Why Civilizations Fail*. North Charleston: CreateSpace Independent Publishing Platform, 2012. (www.rozsavage.com/the-ecological-elephant-in-the-room-goes-up-the-creek-without-a-paddle/)

Business Leadership

Jaworski, Joseph. *Synchronicity: The Inner Path of Leadership*. San Francisco: Berrett-Koehler, 2011.

Laloux, Frederic. *Reinventing Organizations: A Guide to Creating Organizations Inspired by the Next Stage in Human Consciousness*. Brussels: Nelson Parker, 2014.

McChrystal, General Stanley with Tatum Collins, David Silverman and Chris Fussell. *Team of Teams: New Rules of Engagement for a Complex World*. London: Penguin, 2015. (www.rozsavage.com/team-of-teams/)

Wheatley, Margaret J. *Finding Our Way: Leadership for an Uncertain Time*. San Francisco: Berrett-Koehler, 2007. (www.rozsavage.com/finding-our-way/)

Zohar, Danah. *Rewiring the Corporate Brain: Using the New Science to Rethink How We Structure and Lead Organizations*. San Francisco: Berrett-Koehler, 1997.

Social Change

Eisenstein, Charles. *The More Beautiful World Our Hearts Know Is Possible*. Berkeley: North Atlantic Books, 2013.

Evans, Alex. *The Myth Gap: What Happens When Evidence and Arguments Aren't Enough*. London: Eden Project Books, 2017. (www.rozsavage.com/narratives-neuroscience-and-no-money-favourite-books-of-2017/)

Klein, Naomi. *The Shock Doctrine: The Rise of Disaster Capitalism*. London: Penguin, 2008. (www.rozsavage.com/the-shock-doctrine/)

Robinson, Kim Stanley. *The Ministry for the Future*. London: Orbit, 2021. (Fiction)

Rose, Chris. *How to Win Campaigns: Communications for Change*. 2nd ed. Abingdon: Earthscan, 2010.

Sun, Rivera. *The Dandelion Insurrection: Love and Revolution*. El Prado: Rising Sun Press Works, 2013. (Fiction)

Watkins, Alan, and Ken Wilber. *Wicked and Wise: How to Solve the World's Toughest Problems*. 2nd ed. Romsey: Complete Coherence, 2021. (www.rozsavage.com/wicked-and-wise/)

Acknowledgements

There are countless people who over the last twenty years have contributed to my voyages of discovery, both literal and literary. You are giants, and I stand upon your shoulders. Thank you.

I would particularly like to thank Jo de Vries, my editor, who first suggested that I submit a proposal to Flint Books and has been a true champion of this project from start to finish. This book may possibly have existed without her, but it would have been a lot longer and even less comprehensible. Thank you, Jo, for the mercy killing of my darlings. Clint Evans, Ann Luskey and Jo Noble, thank you for being my intrepid beta readers. It's always challenging to give constructive feedback, and you did it with tact and grace, making this a much better book.

Gratitude and appreciation to my stellar podcast guests of 2021: Charles Eisenstein, Paul Hawken, Sharon Blackie, Jude Currivan, Richard Bartlett, Tim Jackson, Kim Stanley Robinson, Kimberly Carter Gamble, Peggy Liu, Bill McKibben, Ted Rau, John Buck and Monika Megyesi. I truly appreciate your wisdom. Thanks also to the podcast production team at dSTUDIO: Basil, Kim, Gian and Gabriel.

Thanks to Kate Maguire, my doctoral supervisor, for helping me build the strong foundation of research and analysis in the dissertation that has now evolved into this book.

Deepest thanks to my amazing manager, Miriam Staley, for always having my back.

Thank you to Irina Stoica, Drigo des Tombes and the rest of the Samarans, who taught me so much about the joys and the challenges of community and connection. Thanks especially to Emily Lane and the Stream Team, who inspired so much of my thinking about the new paradigm.

I am grateful to Lisa Marshall, my coach and mentor, for her wisdom on the importance of yin/yang balance in our world, and to Iain McGilchrist for his invaluable insights on the left and right hemispheres (and for the whisky). Thanks to Biz and Sue for being my travelling companions, and to Jon Keen and Alex Hatfield for the vision quest that led me to Verdi Wren.

I'd like to thank my soul sisters for the moral support, friendship, fun and willingness to navigate uncharted waters with me. Ellen, thank you for the new perspectives and possibilities you bring into my life. You constantly inspire and enlighten me. Terri, thanks for your loyalty and friendship through thick and thin, marriage and divorce, ups and downs.

I'm grateful for my real world neighbours (human, animal, flower and tree). This valley and this village are my haven. Among the humans, thanks especially to Naomi Wilkinson for the walks and talks, and to our beloved Barry, who left us too soon.

And last but far from least, thank you to my mother, Rita, my dear departed father, Hamer, and my sister, Tanya. Our souls chose to share this mortal journey this time around, and for that I am grateful.

Notes

Chapter 1: Environment

1 Daniel Quinn, *Ishmael: A Novel* (New York: Bantam/Turner Books, 1992), 84, Kindle.
2 See my 2013 book, *Stop Drifting Start Rowing* (London: Hay House, 2013), for the full story of this encounter.

Chapter 2: Economics

1. Fawzi Ibrahim, *Capitalism Versus Planet Earth: An Irreconcilable Conflict* (London: Muswell Press, 2012), location 3212, Kindle.
2. Robert F. Kennedy, 'Remarks at the University of Kansas', transcript of speech delivered at the University of Kansas, 18th March 1968, www.jfklibrary.org/learn/about-jfk/the-kennedy-family/robert-f-kennedy/robert-f-kennedy-speeches/remarks-at-the-university-of-kansas-march-18-1968.
3. Simon Kuznets, 'National Income, 1929–1932'. 73rd US Congress, 2d session, Senate document no. 124, 1934, page 7. https://fraser.stlouisfed.org/title/national-income-1929-1932-971.

4. Robert McNamara, 'Beyond GDP: Key Quotes', European Commission, accessed 2nd July 2022, https://ec.europa.eu/environment/beyond_gdp/key_quotes_en.html.
5. Kate Raworth, *Doughnut Economics: Seven Ways to Think Like a 21st-Century Economist* (London: Random House Business, 2017), 27.
6. John Perkins, *Confessions of an Economic Hit Man* (San Francisco: Berrett-Koehler Publishers, 2004), 28.
7. Tim Jackson, *Post Growth: Life After Capitalism* (Cambridge: Polity Press, 2021), 150.
8. 'The Power of Incentives: The Hidden Forces That Shape Behavior', Farnam Street, accessed 2nd July 2022, https://fs.blog/bias-incentives-reinforcement/.
9. John Maynard Keynes, quoted in Michael Albert, *Moving Forward: Programme for a Participatory Economy* (Edinburgh: AK Press, 2000), 128.
10. Max Roser and Esteban Ortiz-Ospina, 'Global Extreme Poverty', Our World in Data, 2013, https://ourworldindata.org/extreme-poverty.
11. Jamie Ducharme, 'This Is the Amount of Money You Need to Be Happy, According to Research', *Money*, 14th February 2018, https://money.com/ideal-income-study/.
12. Emmie Martin, 'Here's How Much Money You Need to be Happy, According to a New Analysis by Wealth Experts', *CNBC*, 20th November 2017, www.cnbc.com/2017/11/20/how-much-money-you-need-to-be-happy-according-to-wealth-experts.html.
13. Julie Miller, 'The Enigma of J. Paul Getty, the One-Time Richest Man in the World', *Vanity Fair*, 22nd December 2017, www.vanityfair.com/hollywood/2017/12/all-the-money-in-the-world-j-paul-getty.
14. Philip Brickman, Dan Coates and Ronnie Janoff-Bulman, 'Lottery Winners and Accident Victims: Is Happiness Relative?', *Journal of Personality and Social Psychology* 36, no. 8 (1978), https://doi.org/10.1037/0022-3514.36.8.917.
15. Teresa Belton, *Happier People Healthier Planet: How Putting Wellbeing First Would Help Sustain Life on Earth* (Bristol: Silverwood Books, 2014), location 38, Kindle.
16. Tim Jackson, 'An Economic Reality Check', TED Global, October 2010, TED Talk video, 6:39–6:58, www.ted.com/talks/tim_jackson_an_economic_reality_check.
17. Edward Bernays, *Propaganda* (New York: Routledge, 1928), 9.

18. John Perkins, *Confessions of an Economic Hit Man* (San Francisco, CA: Berrett-Koehler Publishers Inc., 2004), Preface. Reprinted with permission of the publisher. All rights reserved.
19. Bernard Lietaer and Jacqui Dunne, *Rethinking Money: How New Currencies Turn Scarcity Into Prosperity* (San Francisco: Berrett-Koehler Publishers Inc., 2013), 2. Reprinted with permission of the publisher. All rights reserved.
20. L. Randall Wray, *Understanding Modern Money: The Key to Full Employment and Price Stability* (Cheltenham: Edward Elgar, 1998), viii–ix, quoted in Bernard Lietaer and Jacqui Dunne, *Rethinking Money: How New Currencies Turn Scarcity Into Prosperity* (San Francisco: Berrett-Koehler Publishers, 2013), 26.
21. Quoted by Dennis R. Klinck. 'Pound, Social Credit, and the Critics.' *Paideuma*, vol. 5, no. 2, 1976, pp. 227–40. *JSTOR*, http://www.jstor.org/stable/24725642, accessed 2nd July 2022.
22. John Perkins, *Confessions of an Economic Hit Man* (San Francisco: Berrett-Koehler Publishers, 2004). Reprinted with permission of the publisher. All rights reserved.
23. Before April 2018, Eswatini was known as Swaziland.
24. 'If you ask me what is the worst thing in the world, I will say it is compound interest', Economic Sociology & Political Economy, posted by Oleg Komlik, 21st March 2017, https://economicsociology.org/2017/03/21/the-burden-of-nations-debt-and-compound-interest/if-you-ask-me-what-is-the-worst-thing-in-the-world-i-will-say-it-is-compound-interest/.
25. Warren Buffett, 'Warren Buffett Shares the Secrets to Wealth in America', *Time*, 4th January 2018, https://time.com/5087360/warren-buffett-shares-the-secrets-to-wealth-in-america/.
26. Robert Frank, '"Only morons pay the estate tax," says White House's Gary Cohn', CNBC, 29th August 2017, www.cnbc.com/2017/08/29/only-morons-pay-the-estate-tax-says-white-houses-gary-cohn.html.
27. Thomas Piketty, *Capital in the Twenty-first Century*, trans. Arthur Goldhammer (Cambridge, US: The Belknap Press of Harvard University Press, 2014), location 163, Kindle.
28. Mariana Mazzucato, 'Capitalism After the Pandemic: Getting the Recovery Right', *Foreign Affairs*, November/December 2020, www.foreignaffairs.com/articles/united-states/2020-10-02/capitalism-after-covid-19-pandemic.

29. Richard Wilkinson and Kate Pickett, 'Inequality breeds stress and anxiety. No wonder so many Britons are suffering', *The Guardian*, 10th June 2018, www.theguardian.com/commentisfree/2018/jun/10/inequality-stress-anxiety-britons.
30. Ibid.
31. John de Graaf, David Wann and Thomas H. Naylor, *Affluenza: The All-Consuming Epidemic* (London: McGraw Hill, 2002), Introduction.
32. Paul Gilding, *The great disruption: How the Climate Crisis Will Transform the Global Economy* (London: Bloomsbury, 2011).

Chapter 3: Politics

1. Annie Leonard, *Story of Stuff, Referenced and Annotated Script*, uploaded January 2020, www .storyofstuff.org/wp-content/uploads/2020/01/StoryofStuff_AnnotatedScript.pdf. (Free Press)
2. 'Election 2015: SNP wins 56 of 59 seats in Scots landslide', BBC News. 8th May 2015.
3. 'General Election 2019: Turning votes into seats', House of Commons Library, accessed 2nd July 2022, https://commonslibrary.parliament.uk/general-election-2019-turning-votes-into-seats/.
4. Martin Gilens and Benjamin I. Page, 'Testing Theories of American Politics: Elites, Interest Groups, and Average Citizens', *Perspectives on Politics* 12, no. 3 (2014): 564–81, https://doi.org/10.1017/S1537592714001595, quoted in Andrew Prokop, 'Study: Politicians Listen to Rich People, Not You', *Vox*, 28th January 2015, www.vox.com/2014/4/18/5624310/martin-gilens-testing-theories-of-american-politics-explained.
5. Niall McCarthy, 'How Much Does Money Matter In U.S. Presidential Elections? [Infographic]', *Forbes*, 28th July 2016, www.forbes.com/sites/niallmccarthy/2016/07/28/how-much-does-money-matter-in-u-s-presidential-elections-infographic/.
6. Jan Hanska, *Reagan's Mythical America: Storytelling as Political Leadership* (London: Palgrave Macmillan, 2012), 94.
7. Bill Allison and Sarah Harkins, 'Fixed Fortunes: Biggest Corporate Political Interests Spend Billions, Get Trillions', Sunlight Foundation,

17th November 2014, https://sunlightfoundation.com/2014/11/17/fixed-fortunes-biggest-corporate-political-interests-spend-billions-get-trillions/.
8. Gautham G. Vadakkepatt, Sandeep Arora, Kelly D. Martin and Neeru Paharia, 'Shedding Light on the Dark Side of Firm Lobbying: A Customer Perspective', *Journal of Marketing* 86, no. 3 (May 2022) 79–97, https://doi.org/10.1177/00222429211023040.
9. Seth Thévoz, 'David Cameron Then and Now: How the Ex-PM Changed his Tune on Lobbying', Open Democracy, 12th April 2021, www.opendemocracy.net/en/opendemocracyuk/david-cameron-then-and-now-how-ex-pm-changed-his-tune-lobbying/.
10. Daniel Kreps, 'Jimmy Carter: U.S. Is an "Oligarchy With Unlimited Political Bribery"', *Rolling Stone*, 31st July 2015, www.rollingstone.com/politics/politics-news/jimmy-carter-u-s-is-an-oligarchy-with-unlimited-political-bribery-63262/.
11. Media Reform Coalition, 'Report: Who Owns the UK Media?', 14th March 2021, www.mediareform.org.uk/media-ownership/who-owns-the-uk-media.
12. 'Malcolm X on racism, capitalism and Islam', Aljazeera, 21st February 2022, accessed 2nd July 2022, www.aljazeera.com/news/2022/2/21/malcolm-x-quotes.
13. Jonathan Capehart, 'Republicans had it in for Obama Before Day 1', *The Washington Post*, 10th August 2012, www.washingtonpost.com/blogs/post-partisan/post/republicans-had-it-in-for-obama-before-day-1/2012/08/10/0c96c7c8-e31f-11e1-ae7f-d2a13e249eb2_blog.html.
14. Albert Mehrabian, *Silent Messages* (1st ed.), (Belmont, CA: Wadsworth, 1971).
15. Charles C. Ballew II and Alexander Todorov, 'Predicting political elections from rapid and unreflective face judgments', *Proceedings of the National Academy of Sciences of the United States of America* 104, no. 46 (2007): 17948-17953, doi: 10.1073/pnas.0705435104.
16. Estimate as at 12th June 2015, quoted in www.telegraph.co.uk/news/politics/tony-blair/11670425/Revealed-Tony-Blair-worth-a-staggering-60m.html.
17. Dan Alexander, 'How Bill And Hillary Clinton Made $240 Million In The Last 15 Years', *Forbes*, 8th November 2016, www.forbes.com/sites/danalexander/2016/11/08/how-bill-house-hillary-clinton-made-240-million-how-much-earnings-rich-white/?sh=5fdde97b7a16.

18. Mahnoor Khan, 'Putin Claims he Makes $140,000 and has an 800-Square Foot Apartment. His Actual Net Worth is a Mystery No One Can Solve', Fortune, 2nd March 2022, https://fortune.com/2022/03/02/vladimir-putin-net-worth-2022/.
19. Jill Goldsmith, 'WarnerMedia CEO Jason Kilar's 2020 Pay Package Totaled $52 Million With Big Stock Award', *Deadline*, 11th March 2021, https://deadline.com/2021/03/warnermedia-ceo-jason-kilar-2020-pay-totaled-52-million-with-big-stock-award-1234712596/.
20. Video of her DNC speech available at www.youtube.com/watch?v=HzEy_LTgh50. Excerpt at 4:50 mins. Accessed 2nd July 2022.
21. Stephanie Marsh, 'Marianne Williamson: The "Leftwing Trump" Preaching the Politics of Love', *The Observer*, 30th July 2019, www.theguardian.com/global/2019/jul/30/marianne-williamson-can-love-beat-trump-in-the-2020-presidential-elections.
22. Maanvi Singh, '"Politics of love": The End of Marianne Williamson's Bizarre and Mesmerizing Campaign', *The Guardian*, 11th January 2020, www.theguardian.com/us-news/2020/jan/10/marianne-williamson.
23. 'The Most Admired Man and Woman', Gallup, accessed 2nd July 2022, https://news.gallup.com/poll/1678/most-admired-man-woman.aspx.

Chapter 4: Power

1. adrienne maree brown, *Emergent Strategy*. (Edinburgh, Scotland: AK Press, 2017), 197.
2. The video of our conversation is available online: www.youtube.com/watch?v=or8ps4x2tdQ.
3. Riane Eisler, *The Chalice and the Blade: Our History, our Future* (San Francisco: Perennial Library, 1988), xiii.
4. Riane Eisler, *The Chalice and the Blade: Our History, our Future* (San Francisco: Perennial Library, 1988), 154.
5. I'm aware some people regard her analysis as controversial or even plain wrong. I shall leave that debate to the archaeologists. My view is: a) a new analysis is always contested when it upsets the status quo, and b) I'm less interested in Riane's archaeological evidence and more interested in what she has to say about the need to transition from a domination paradigm to one of partnership. This seems

self-evident to me, and I don't need to know if it has existed in the past to know that I want it to exist in the future.

6. 'Lilith: Who is Lilith', Kenyon College, accessed 2nd July 2022, www2.kenyon.edu/Depts/Religion/Projects/Reln91/Power/lilith.htm.
7. Sharon Blackie, *If Women Rose Rooted: The Journey to Authenticity and Belonging* (Tewksbury: September Publishing, 2016), location 110, Kindle. Reprinted with permission.
8. Ibid., 122, Kindle. Reprinted with permission.
9. Genesis 1:26 (English Standard Version).
10. Liz Jakimow, 'Genesis 1:26–28 and Environmental Rights', Right Now, 22nd February 2013, https://rightnow.org.au/opinion-3/genesis-126-28-and-environmental-rights/.
11. Sharon Blackie, *If Women Rose Rooted: The Journey to Authenticity and Belonging* (Tewksbury: September Publishing, 2016), Kindle. Reprinted with permission.
12. Marc Gafni, 'The Wheel of Co-Creation 2.0', accessed 2nd July 2022, www.marcgafni.com/the-wheel-of-co-creation-2-0/.
13. Daniel Quinn, *Ishmael: A Novel* (New York: Bantam/Turner Books, 1992), 132, Kindle.
14. 'World Hunger Facts', Action Against Hunger, accessed 2nd July 2022, www.actionagainsthunger.org.uk/why-hunger/world-hunger-facts.
15. Nicolas Vega, 'Elon Musk Says he is Willing to Spend $6 billion to Fight World Hunger – On One Condition', NBC News, 2nd November 2021, www.nbcnews.com/business/business-news/elon-musk-says-willing-spend-6-billion-fight-world-hunger-one-conditio-rcna4301.
16. 'Rising Inequality Affecting More Than Two-Thirds of the Globe, But it's Not Inevitable: New UN Report', United Nations, 21st January 2020, https://news.un.org/en/story/2020/01/1055681.
17. Lawrence Mishel and Jori Kandra, 'CEO pay has Skyrocketed 1,322% Since 1978', Economic Policy Institute, 10th August 2021, www.epi.org/publication/ceo-pay-in-2020/.
18. Michael Sandel, *The Tyranny of Merit: What's Become of the Common Good?* (London: Allen Lane, 2020), location 803, Kindle.
19. Alain de Botton, 'A Kinder, Gentler Philosophy of Success', TED, July 2009, TED Talk video, 6:40–7:14, www.ted.com/talks/alain_de_botton_a_kinder_gentler_philosophy_of_success.
20. Joe Pinsker, 'The Financial Perks of Being Tall', *The Atlantic*, 18th May 2015, www.theatlantic.com/business/archive/2015/05/the-financial-perks-of-being-tall/393518/.

21. Cilo Andris, David Lee, Marcus J. Hamilton, Mauro Martino, *et al.*, 'The Rise of Partisanship and Super-Cooperators in the U.S. House of Representatives', *PLoS ONE* 10, no. 4 (21st April 2015): e0123507, https://doi.org/10.1371/journal.pone.0123507.
22. Ken Robinson, 'Do Schools Kill Creativity?', TED, 27th June 2006, TED Talk video, 11:34–11:59, www.ted.com/talks/sir_ken_robinson_do_schools_kill_creativity/transcript?language=en.
23. Sugata Mitra, *The Hole in the Wall Project and the Power of Self-Organized Learning*, www.edutopia.org/blog/self-organized-learning-sugata-mitra.
24. Cathy O'Neil, *Weapons of Math Destruction: How Big Data Increases Inequality and Threatens Democracy* (London: Penguin, 2017), 117, Kindle.
25. Tristan Harris, 'Humane: A New Agenda for Tech with Tristan Harris – Presentation and Transcript', transcript of presentation delivered at Humane: A New Agenda for Tech, held at SFJAZZ Center in San Francisco on 23rd April 2019, Ethical, 5th May 2019, https://ethical.net/ethical/humane-new-agenda-tech-tristan-harris/.
26. Podcast interview with Charles Eisenstein is available at https://podcasts.apple.com/us/podcast/charles-eisenstein-from-separation-to-interbeing/id1591297431?i=1000539214696.
27. Darrell Trent, quoted in Andri Stavrou, 'Defining Terrorism: A Matter of Perspective', European Student Think Tank, 15th April 2019, https://esthinktank.com/2019/04/15/defining-terrorism-a-matter-of-perspective/.
28. Podcast interview with Charles Eisenstein is available at https://podcasts.apple.com/us/podcast/charles-eisenstein-from-separation-to-interbeing/id1591297431?i=1000539214696.
29. Noam Chomsky, *The Common Good*, interview by David Barsamian, ed. Arthur Naiman (Monroe: Odonian Press, 1998).
30. Hannah Arendt, 'Hannah Arendt: From an Interview', *The New York Review*, 26th October 1978, www.nybooks.com/articles/1978/10/26/hannah-arendt-from-an-interview/.
31. Jon Jureidini and Leemon B. McHenry, 'The Illusion of Evidence Based Medicine' *BMJ* 376, no. 0702 (16th March 2022), https://doi.org/10.1136/bmj.o702.
32. For reasons to be sceptical about the pharmaceutical industry, see, for example: Laura Spinney, 'Covid Vaccines Deserve our Trust – but Big Pharma Doesn't', *The Guardian*, 9th February 2022, www.

theguardian.com/commentisfree/2022/feb/09/covid-vaccines-lifesavers-big-pharma-mistrust, and Stephen Buranyi, 'Big Pharma is Fooling Us', *The New York Times*, 17th December 2022, www.nytimes.com/2020/12/17/opinion/covid-vaccine-big-pharma.html.
33. Alan Watts, *The Book: On the Taboo Against Knowing Who You Are* (London: Souvenir Press, 2009), 10.
34. Yuval Noah Harari, *21 Lessons for the 21st Century* (London: Vintage, 2018), 223, Kindle.
35. Alice Walker, quoted in William P. Martin, *The Best Liberal Quotes Ever: Why the Left is Right* (Naperville: Sourcebooks, 2004), 173.
36. Mary Beard, *Women & Power: A Manifesto* (London: Profile Books, 2017), 54, Kindle.
37. 'The verb *handbag* means, especially of a woman, to bully or coerce by subjecting to a forthright verbal assault or criticism' quoted in, Pascal Treguer, '"Handbag": How Thatcher Enriched The English Language', World Histories, accessed 2nd July 2022, https://wordhistories.net/2018/03/16/handbag-margaret-thatcher/.
38. As recounted by Sheryl Sandberg in *Lean In: Women, Work, and the Will to Lead* (London, Ebury Publishing/RandomHouse, 2013), location 583, Kindle.
39. Mary Beard, *Women & Power: A Manifesto* (London: Profile Books, 2017), 61, Kindle.
40. Podcast interview with Tim Jackson is available at https://podcasts.apple.com/us/podcast/professor-tim-jackson-happiness-and-wellbeing-in-a/id1591297431?i=1000539217982.
41. Juliet Diaz, 'What We're Getting Wrong About Modern Spirituality, From A Spiritual Activist', interview by Sarah Regan, Mind Body Green, 8th March 2022, https://amp.mindbodygreen.com/articles/the-problem-with-modern-spirituality-follow-her-lead.

Chapter 5: Psychology

1. Aldous Huxley, *Island* (first published in 1962 by Chatto & Windus, London: Vintage/RandomHouse, 2005), 190, Kindle.
2. Daniel Simons and Christopher Chabris, 'Selective Attention Test', Danial Simons, 10th March 2010, YouTube video, 1:21, www.youtube.com/watch?v=vJG698U2Mvo.

3. If you didn't see the gorilla, watch this video to make yourself feel better: Daniel Simons and Daniel Levin, 'The "Door" Study', Daniel Simons, 14th March 2010, YouTube video, 1:36, www.youtube.com/watch?v=FWSxSQsspiQ.
4. Dylan Wiliam, 'The Half-Second Delay: What Follows?', *Pedagogy, Culture and Society* 14, no. 1 (22nd August 2006), 71–81, https://doi.org/10.1080/14681360500487470.
5. Alison Gopnik, *The Philosophical Baby: What Children's Minds Tell Us about Truth, Love and the Meaning of Life* (London: Bodley Head, 2009) and http://alisongopnik.com/lantern_v_spotlight.htm.
6. Adam Rutherford, 'Well You Would Say That: The Science Behind Our Everyday Biases', *The Observer*, 16th October 2021, www.theguardian.com/science/2021/oct/16/the-science-behind-our-everyday-biases-confirmation-polarisation-psychology-dunning-kruger.
7. Surveys conducted by the YPCCC discovered that, rather than a continuous spectrum, there are six distinct demographics in the US public when it comes to climate change, ranging from 'Alarmed' to 'Dismissive'. Each group has different psychological, cultural and political reasons for acting – or not acting – on climate change. See: 'Global Warming's Six Americas', Yale Program on Climate Change Communication, https://climatecommunication.yale.edu/about/projects/global-warmings-six-americas/.
8. Iain McGilchrist, *The Master and His Emissary: The Divided Brain and the Making of the Western World* (London: Yale University Press, 2009), 97.
9. Larry Dossey, *How Our Individual Mind is Part of a Greater Consciousness and Why it Matters* (Carlsbad, CA: Hay House, 2013), 207, Kindle.
10. Ibid., 209.
11. This, and many more hilarious examples here: Dina Rickman, 'Bizarre Correlations That Will Leave You Wishing Nicolas Cage Would Retire', *Indy100*, 10th August 2014, www.indy100.com/offbeat/bizarre-correlations-that-will-leave-you-wishing-nicolas-cage-would-retire-7240456. The divorce rate in Maine correlates with per capita margarine consumption, global warming is caused by a pirate shortage, etc.
12. Fun probabilities and improbabilities here: Pandora49, 'What's the Most Unlikely Thing That Can Happen to You? (RLL #1)', JetPunk, last modified 7th April 2021, www.jetpunk.com/users/sacheth9/blog/whats-the-most-unlikely-thing-that-can-happen-to-you.

13. C.A. Hallmann; M. Sorg; E. Jongejans; H. Siepel; N. Hofland; H. Schwan (18th October 2017), 'More than 75 percent decline over 27 years in total flying insect biomass in protected areas', *PLOS ONE*, 12 (10): e0185809, doi:10.1371/journal.pone.0185809.
14. Dan Gilbert, 'The Psychology of Your Future Self', TED, June 2014, TED Talk video, 00:05–00:28, www.ted.com/talks/dan_gilbert_the_psychology_of_your_future_self?language=en.
15. Martin Niemöller, 'First They Came', Holocaust Encyclopedia, United States Holocaust Memorial Museum, last modified 30th March 2012, https://encyclopedia.ushmm.org/content/en/article/martin-niemoeller-first-they-came-for-the-socialists.
16. Stanley Milgram, *The Perils of Obedience* (Harper's, 1973), 62, accessed 2nd July 2022, https://is.muni.cz/el/1423/podzim2013/PSY268/um/43422262/Milgram_-_perils_of_obediance.pdf.
17. Ibid., 76.
18. 'Authority Bias: Lessons from the Milgram Obedience Experiment', *Effectiviology*, accessed 2nd July 2022, https://effectiviology.com/authority-bias-the-milgram-obedience-experiment/.
19. Ariella S. Kristal and Laurie R. Santos, 'G.I. Joe Phenomena: Understanding the Limits of Metacognitive Awareness on Debiasing' (Harvard Business School Working Paper, No. 21–084, January 2021), 54, www.hbs.edu/faculty/Pages/item.aspx?num=59722.
20. 'Culture Eats Strategy for Breakfast', Quote Investigator, modified on 25th May 2017, https://quoteinvestigator.com/2017/05/23/culture-eats/.
21. For example: Elliot T. Berkman, 'The Neuroscience of Goals and Behaviour Change', *Consulting Psychology Journal: Practice and Research* 70, no. 1 (2018), 28–44, https://doi.org/10.1037/cpb0000094. For an overview: Elliot T. Berkman, 'Why Is Behavior Change So Hard?', *Psychology Today*, 20th March 2018, www.psychologytoday.com/us/blog/the-motivated-brain/201803/why-is-behavior-change-so-hard.

Chapter 6: Hemispheres

1. Iain McGilchrist, *The Master and His Emissary: The Divided Brain and the Making of the Western World* (London: Yale University Press, 2009), 460.

2. Edmund Newell, *The Sacramental Sea: A Spiritual Voyage Through Christian History* (London: Darton, Longman and Todd, 2019).
3. Margaret Heffernan, *Wilful Blindness: Why We Ignore the Obvious* (London: Simon & Schuster, 2012).
4. Iain McGilchrist, *The Master and His Emissary: The Divided Brain and the Making of the Western World* (London: Yale University Press, 2012).
5. Iain McGilchrist, *The Divided Brain and the Search for Meaning* (London: Yale University Press, 2012). There is an excellent summary of Iain's analysis in a short animated film in association with the RSA: Abi Stephenson and Cognitive Media, 'RSA Animate: The Divided Brain', RSA, 21st October 2011, YouTube video, 11:47, www.youtube.com/watch?v=dFs9WO2B8uI.
6. Iain McGilchrist, *The Master and his Emissary: The Divided Brain and the Making of the Western World* (London: Yale University Press, 2009), location 519, Kindle.
7. Weiwei Men, *et al.*, 'The Corpus Callosum of Albert Einstein's Brain: Another Clue to his High Intelligence?', *Brain* 137, no. 4 (24th September 2013), e268, https://doi.org/10.1093/brain/awt252.
8. Manfredo Massironi, *The Psychology of Graphic Images: Seeing, Drawing, Communicating* (London: Psychology Press, 2001), 163.
9. Sam Shead, '*Silicon Valley's quest to live forever could benefit humanity as a whole — here's why*', CNBC, 21st September 2021, www.cnbc.com/2021/09/21/silicon-valleys-quest-to-live-forever-could-benefit-the-rest-of-us.html.
10. Melissa Dahl, 'Huh, Would You Believe That Forcing Employees to Act Happy Is a Bad Idea?', *The Cut*, 7th November 2016, www.thecut.com/2016/11/forcing-employees-to-act-happy-is-a-terrible-idea.html.
11. Cathy O'Neil, *Weapons of Math Destruction: How Big Data Increases Inequality and Threatens Democracy* (London: Penguin, 2017).
12. Christian Jarrett, 'Psychology: How Many Senses do we Have?', *BBC Future*, 19th November 2014, www.bbc.com/future/article/20141118-how-many-senses-do-you-have.
13. Susan J. Rasmussen, 'Only Women Know Trees: Medicine Women and the Role of Herbal Healing in Tuareg Culture', *Journal of Anthropological Research* 54, no. 2 (Summer 1998), 147–71, http://www.jstor.org/stable/3631728.
14. Paul Kalanithi, *When Breath Becomes Air* (London: Vintage, 2017).

15. George Markowsky, 'Physiology', *Britannica*, last modified 16th June 2017, www.britannica.com/science/information-theory/Physiology.
16. Iain McGilchrist, *The Divided Brain and the Search for Meaning* (London: Yale University Press, 2012), location 420, Kindle.
17. Robert Logan, *The Alphabet Effect: A Media Ecology Understanding of Western Civilization* (New York: Hampton Press, 2004).

Chapter 7: Systems

1. Donella H. Meadows, 'Whole Earth Models and Systems', *The CoEvolution Quarterly* (Summer 1982), 98–108, https://donellameadows.org/wp-content/userfiles/Whole-Earth-Models-and-Systems.pdf.
2. But more likely originating with Henry Thomas Buckle: see https://quoteinvestigator.com/2014/11/18/great-minds/, accessed 2nd July 2022.
3. Donella H. Meadows, Dennis L. Meadows, Jørgen Randers and William W. Behrens III, *The Limits to Growth* (New York: Universe Books, 1972), available as a pdf at http://www.donellameadows.org/wp-content/userfiles/Limits-to-Growth-digital-scan-version.pdf.
4. Donella Meadows, 'Leverage Points: Places to Intervene in a System', The Donella Meadows Project: The Academy for Systems Change, accessed 2nd July 2022, https://donellameadows.org/archives/leverage-points-places-to-intervene-in-a-system/.
5. Ibid.
6. Ibid.
7. Richard Heinberg, 'Systemic Change Driven by Moral Awakening Is Our Only Hope', *EcoWatch*, 14th August 2017, www.ecowatch.com/climate-change-heinberg-2471869927.html.
8. Stockholm Resilience Centre, 'Planetary Boundaries', Stockholm University, accessed on 2nd July 2022, www.stockholmresilience.org/research/planetary-boundaries.html.
9. Ralitsa Vassileva, 'Thomas Friedman: We Need Greedy Leaders', *Sustainia*, 29th May 2018, https://sustainiaworld.com/only-greed-can-make-the-world-sustainable.
10. Tim Holmes, Elena Blackmore, Richard Hawkins and Tom Wakeford, *The Common Cause Handbook* (Machynlleth: Public

Interest Research Centre, 2011), https://commoncausefoundation.org/wp-content/uploads/2021/10/CCF_report_common_cause_handbook.pdf.

11. Chris Rose, *How to Win Campaigns: 100 Steps to Success* (London: Earthscan, 2005), 39. See also Chris Rose's website, http://www.campaignstrategy.org/, and the Cultural Dynamics website, http://www.cultdyn.co.uk/, for invaluable resources on cultural change.
12. Daniel Kahneman, 'The Riddle of Experience vs. Memory', TED, March 2010, Ted Talk video, 19:50, www.ted.com/talks/daniel_kahneman_the_riddle_of_experience_vs_memory.
13. Quoted by David Peter Stroh in *Systems Thinking For Social Change: A Practical Guide to Solving Complex Problems, Avoiding Unintended Consequences, and Achieving Lasting Results* (London: Chelsea Green Publishing, 2015), 257, Kindle.

Chapter 8: Collapse

1. Pema Chödrön, *When Things Fall Apart: Heart Advice for Difficult Times* (London: Element, 2005), 81.
2. William Hutton, *Coming Earth Changes: Causes and Consequences of the Approaching Pole Shift* (Virginia Beech: ARE Press, 1996). Original source material available directly from the Edgar Cayce Foundation via https://dokument.pub/21074-earth-changes-flipbook-pdf.html: *Earth Changes: A compilation of Extracts from the Edgar Cayce Readings*, 24. Edgar Cayce Readings © 1971, 1993–2007, The Edgar Cayce Foundation. All Rights Reserved.
3. Ibid., 36.
4. Thomas Mails, *The Hopi Survival Kit: The Prophecies, Instructions and Warnings Revealed by the Last Elders* (London: Penguin, 1997).
5. Raymond Williams, *Resources of Hope* (London: Verso, 1989), 118.
6. Adam Dorr, 'Common Errors in Reasoning About the Future: Three Informal Fallacies', *Technological Forecasting and Social Change* 116 (March 2017), 322–30, https://doi.org/10.1016/j.techfore.2016.06.018.
7. William Ophuls, *Immoderate Greatness: Why Civilizations Fail* (North Charleston: CreateSpace Independent Publishing Platform, 2012).

8. Owen Mulhern, 'The Time Lag of Climate Change', Earth.Org, 27th August 2020, https://earth.org/data_visualization/the-time-lag-of-climate-change.
9. Tracey Peake, 'Time Lag Between Intervention and Actual CO_2 Decrease Could Still Lead to Climate Tipping Point', New York State University, 15th December 2021, https://news.ncsu.edu/2021/12/time-lag-could-still-lead-to-climate-tipping-point/.
10. See, for example, *United Nations, Department of Economic and Social Affairs, Population Division (2019). World Population Prospects 2019: Ten Key Findings.* Other reports from the United Nations on Population Dynamics at https://population.un.org/wpp/Publications/.
11. 'About', Earth Overshoot Day, accessed 2nd July 2022, www.overshootday.org/about/.
12. Earth Overshoot Day is recalculated every year for all past years, as well as the current year, to use the latest reported data and science. The dates quoted here are as calculated at the time of writing, in 2021.
13. Robert Peston, *How Do We Fix This Mess? The Economic Price of Having it all, and the Route to Lasting Prosperity* (London: Hodder, 2013), 14.
14. John Bagot Glubb, *The Fate of Empires and Search for Survival* (Edinburgh: Blackwood, 1978).
15. William Ophuls, *Immoderate Greatness: Why Civilizations Fail* (Scotts Valley, CA: CreateSpace Independent Publishing Platform, 2012), 49.
16. Jared Diamond, 'Why do societies collapse?', TED Global, October 2008, Ted Talk Video, www.ted.com/talks/jared_diamond_why_do_societies_collapse.
17. J. Diamond, *Collapse: How Societies Choose to Fail or Survive* (Viking Press, 2005).
18. Cryonics Institute, www.cryonics.org/.
19. 'Elon Musk Puts a Cap on Ticket Price to Mars Colony. Here's how much it could cost', Futurism, 15 June 2017, https://futurism.com/elon-musk-puts-a-cap-on-ticket-price-to-mars-colony-heres-how-much-it-could-cost.
20. John Kenneth Galbraith, *The Age of Uncertainty* (1977), Chapter 1, 22.
21. William Ophuls, *Immoderate Greatness: Why Civilizations Fail* (Scotts Valley, CA: CreateSpace Independent Publishing Platform, 2012), 2.
22. Ibid., 54.
23. Arnold J. Toynbee, *A Study of History: Abridgement of Volumes I to VI* (Oxford: Oxford University Press, 1947).

Chapter 9: Paradigms

1. Sharon Blackie, *If Women Rose Rooted: The Journey to Authenticity and Belonging* (Tewksbury: September Publishing, 2019), location 1005, Kindle. Reprinted with permission.
2. Hopi Elders, 'We are the Ones We've Been Waiting For: Prophecy made by Hopi Elders', University of Minnesota, 23rd July 2020, https://artistic.umn.edu/we-are-ones-weve-been-waiting-prophecy-made-hopi-elders.
3. Barbara Marx Hubbard, *Conscious Evolution: Awakening the Power of our Social Potential* (Novato, CA: New World Library, 1998), location 3134, Kindle.
4. Viktor Frankl, *Man's Search for Meaning* (London: Penguin Random House, 2020), 139, Kindle.
5. Richard Rohr, *Spiral of Violence: The World, the Flesh, and the Devil* (CAC: 2008), audio.
6. David Foster Wallace, 'This Is Water by David Foster Wallace Full Speech', uploaded by Joe Mita, 5th May 2013, YouTube video, 0:17–0:35, www.youtube.com/watch?v=PhhC_N6Bm_s.
7. Timothy Morton, *Hyperobjects: Philosophy and Ecology After the End of the World* (Minneapolis: University of Minnesota Press, 2013).
8. Yuval Noah Harari, *21 Lessons for the 21st Century* (London: Vintage, 2018), 271, Kindle.
9. 'Acupuncture', Wikipedia, accessed 7th July 2022, https://en.wikipedia.org/wiki/Acupuncture.
10. TED Staff, 'Open for Discussion: Graham Hancock and Rupert Sheldrake from TEDxWhitechapel', TED Blog, 14th March 2013, https://blog.ted.com/open-for-discussion-graham-hancock-and-rupert-sheldrake/.
11. 'Defense.gov News Transcript: DoD News Briefing – Secretary Rumsfeld and Gen. Myers', United States Department of Defense (defense.gov), 12th February 2022, accessed 7th July 2022, https://archive.ph/20180320091111/http:/archive.defense.gov/Transcripts/Transcript.aspx?TranscriptID=2636.
12. Somewhat debunked here: 'Counting Thoughts, Part I', Exploring the Problem Space, 1st January 2017, www.exploringtheproblemspace.com/new-blog/2017/1/1/counting-thoughts-part-i.

13. Jason Murdock, 'Humans Have More than 6,000 Thoughts per Day, Psychologists Discover', *Newsweek*, 15th July 2020, www.newsweek.com/humans-6000-thoughts-every-day-1517963.
14. 'Meet the Founder: Louise LeBrun', The WEL-Systems Institute, accessed 7th July 2022, https://wel-systems.com/meet-the-founder/.
15. Deborah Byrd, 'How and When did the First Planets form in our Universe?', *EarthSky*, 15th September 2012, https://earthsky.org/space/how-and-when-did-the-first-planets-form-in-our-universe/.
16. Itzhak Bars and John Terning, *Extra Dimensions in Space and Time* (Springer, 2010), 27, accessed 7th July 2022.
17. 'How many stars are there in the Universe', The European Space Agency, accessed 7th July 2022, www.esa.int/Science_Exploration/Space_Science/Herschel/How_many_stars_are_there_in_the_Universe.
18. Anthony Robinson, 'How Many Planets in the Universe?', Skies and Scopes, accessed 7th July 2022, https://skiesandscopes.com/how-many-planets/.
19. 'Dark Energy, Dark Matter', NASA Science, accessed 7th July 2022, https://science.nasa.gov/astrophysics/focus-areas/what-is-dark-energy.
20. Dave Gray, *Liminal Thinking: Create the Change You Want by Changing the Way You Think* (New York: Two Waves Books, 2016), location 658, Kindle.
21. Maria Popova, 'How Mendeleev Invented His Periodic Table in a Dream', *The Marginalian*, 8th February 2016, www.themarginalian.org/2016/02/08/mendeleev-periodic-table-dream.
22. Arne Klingenberg, *Beyond Machine Man: Who we really are and why Transhumanism is just an empty promise!* (Port Douglas: Beam Publishing, 2021).

Chapter 10: Chaos

1. Rivera Sun, *Winds of Change* (El Prado: Rising Sun Press Works, 2021), 261, Kindle.
2. E.N. Lorenz, 'Predictability: Does the Flap of a Butterfly's Wings in Brazil Set off a Tornado in Texas?', American Association for the Advancement of Science (1972), https://eapsweb.mit.edu/sites/default/files/Butterfly_1972.pdf, accessed 7th July 2022.

3. Alan Watkins and Ken Wilber, *Wicked and Wise: How to Solve the World's Toughest Problems* (Romsey: Complete Coherence, 2015), location 402, Kindle.
4. Scott Page, quoted in William Ophuls, *Immoderate Greatness: Why Civilizations Fail* (North Charleston: CreateSpace Independent Publishing Platform, 2012), 37.
5. Jack O'Connor, 'New UN University Report: Disasters Around the World are Interconnected', United Nations University press release, 8th September 2021, https://unu.edu/media-relations/releases/new-un-university-report-disasters-around-the-world-are-interconnected.html.
6. Jack O'Connor, *et al.*, *Interconnected Disaster Risks* (Bonn: United Nations University – Institute for Environment and Human Security, 2021), 8, https://interconnectedrisks.org/.
7. M. Mitchell Waldrop, *Complexity: The Emerging Science at the Edge of Order and Chaos* (New York: Simon & Schuster, 1992), 231.
8. Reed Hastings, 'Extra: What if Your Company Had No Rules?', 12th September 2020, in *Freakonomics Radio*, produced by Mary Diduch, podcast, MP3 audio, 58:24, https://freakonomics.com/podcast/book-club-hastings/.
9. Sebastian Bathiany, *et al.*, 'Abrupt Climate Change in an Oscillating World', *Scientific Reports* 8, no. 5040 (March 2018), https://doi.org/10.1038/s41598-018-23377-4.
10. Malcolm Gladwell, *The Tipping Point: How Little Things Can Make a Big Difference*, (London: Abacus, 2002), p.259.
11. Everett Rogers, *Diffusion of Innovations, 5th Edition* (London: Simon & Schuster, 2003).
12. Simon Sinek, 'How great leaders inspire action', TEDxPuget Sound, May 2010, TED Talk Video, www.ted.com/talks/simon_sinek_how_great_leaders_inspire_action/transcript, from 10:47 mins.
13. Les Robinson, 'A Summary of Diffusion of Innovations', Changeology, accessed October 2020, http://www.enablingchange.com.au/Summary_Diffusion_Theory.pdf.
14. Ibid.
15. Paul H. Ray, and Sherry Ruth Anderson, *The Cultural Creatives: How 50 Million People Are Changing the World* (New York: Crown Publications, 2001).

16. Glenda H. Eoyang, *Coping with Chaos: Seven Simple Tools* (Cheyenne: Lagumo Corporation, 1997), 110.
17. I have done a lot of work with the SEEDS currency, which is regenerative by design, rather than degenerative by default, to paraphrase Kate Raworth. See https://joinseeds.earth/.
18. For example: 'Mega-rich Recoup COVID-Losses in Record-time yet Billions will Live in Poverty for at Least a Decade', Oxfam International press release, 25th January 2021, www.oxfam.org/en/press-releases/mega-rich-recoup-covid-losses-record-time-yet-billions-will-live-poverty-least.
19. Green America, *Our Interview with Dr Bernard Lietaer*, www.greenamerica.org/show-ga-blog?nid=5585.
20. Reed Hastings, 'Extra: What if Your Company Had No Rules?', 12th September 2020, in *Freakonomics Radio*, produced by Mary Diduch, podcast, MP3 audio, 58:24, https://freakonomics.com/podcast/book-club-hastings/.
21. For example: 'Coding Adventure: Boids', Sebastian Lague, 26th August 2019, YouTube video, 8:34, www.youtube.com/watch?v=bqtqltqcQhw.
22. Jonathan Glancy, James V. Stone and Stuart P. Wilson, 'How Self-organization Can Guide Evolution', *Royal Society Open Science* 3, no. 11 (November 2016): 160553, https://doi.org/10.1098/rsos.160553.
23. Barbara Marx Hubbard, 'Conscious Evolution and the Integration of Science and Spirituality', Great Mystery, accessed 7th July 2022, https://greatmystery.org/barbara-marx-hubbard/.
24. Alan Watkins and Iman Stratenus, *Crowdocracy: The End of Politics* (Romsey: Urbane Publications, 2016).
25. John Kania and Mark Kramer, 'Collective Impact', *Stanford Social Innovation Review 9*, no. 1 (Winter 2011), 36–41, https://doi.org/10.48558/5900-KN19.
26. Frederic Laloux, *Reinventing Organizations: A Guide to Creating Organizations Inspired by the Next Stage in Human Consciousness* (Brussels: Nelson Parker, 2014).
27. General Stanley McChrystal with Tatum Collins, David Silverman and Chris Fussell, *Team of Teams: New Rules of Engagement for a Complex World* (London: Penguin, 2015), location 163, Kindle.
28. Margaret J. Wheatley, *Finding Our Way: Leadership for an Uncertain Time* (San Francisco: Berrett-Koehler Publishers, 2007), 158.

29. Deniz Igan, Prachi Mishra and Thierry Tressel, ‘A Fistful of Dollars: Lobbying and the Financial Crisis’ (working paper, IMF, December 2009) 72, www.imf.org/external/pubs/ft/wp/2009/wp09287.pdf.
30. Note: I’m aware that similar technologies currently being proposed by the World Economic Forum are evoking fears of increased surveillance, commercial exploitation of personal data and social credit-type systems that penalise individuals whose behaviour is judged to be contrary to the public good. I emphasise here that the apps I envisage would vouchsafe confidentiality, with data being used purely for an individual to evaluate their own behaviour. Data would absolutely not be aggregated, traded or monitored. This may be to place too much faith in technology providers, but that is a discussion for another time.
31. A zettabyte being a unit of digital information that is 1,000,000,000,000,000,000,000 bytes or a trillion gigabytes – see: Branka Vuleta, ‘How Much Data Is Created Every Day? [27 Staggering Stats]’, Seed Scientific, 28th October 2021, https://seedscientific.com/how-much-data-is-created-every-day.
32. Margaret Heffernan, ‘Dare to Disagree’, TED, August 2012, TED Talk video, 12:40, www.ted.com/talks/margaret_heffernan_dare_to_disagree.
33. Ibid., 4:36–5:13.
34. Reed Hastings, ‘Extra: What if Your Company Had No Rules?’, 12th September 2020, in *Freakonomics Radio*, produced by Mary Diduch, podcast, MP3 audio, 58:24, https://freakonomics.com/podcast/book-club-hastings/.
35. ‘MIT’s Building 20: “The Magical Incubator” (1998)’, From the Vault of MIT, 20th January 2016, YouTube video, 36:25, https://infinite-history.mit.edu/video/mits-building-20-magical-incubator.
36. Piero Ferrucci, *What We May Be: Techniques for Psychological and Spiritual Growth Through Psychosynthesis* (New York: Jeremy P. Tarcher, 2009), 297.
37. ‘Nearly all FTSE 100 Companies Have met the Parker Review’s “One before 2021” Target to Improve Ethnic Diversity of FTSE 100 Boards’, EY press release, 16th March 2022, www.ey.com/en_uk/news/2022/03/nearly-all-ftse-100-companies-have-met-the-parker-review-s-one-before-2021-target-to-improve-ethnic-diversity-of-ftse-100-boards.

38. Kevin Rawlinson, 'FTSE Firms' Excuses for Lack of Women in Boardrooms "Pitiful and Patronising"', *The Guardian*, 31st May 2018, www.theguardian.com/business/2018/may/31/pitiful-views-on-women-in-boardrooms-permeate-ftse-firms.
39. Daniel Schmachtenberger, 'Daniel Schmachtenberger's talk at Emergence', uploaded by John B, 29th August 2016, YouTube video, 25:06, https://civilizationemerging.com/media-old/emergence/.
40. Andy Coghlan, 'Science : What Really Makes Water Wet?', *New Scientist*, 15th February 1997, www.newscientist.com/article/mg15320693-200-science-what-really-makes-water-wet/.
41. Todd E. Feinberg and Jon Mallatt, 'Phenomenal Consciousness and Emergence: Eliminating the Explanatory Gap', *Frontiers in Psychology* 11 (12th June 2020), 1041, https://doi.org/10.3389/fpsyg.2020.01041.
42. Elizabeth Gilbert, 'Elizabeth Gilbert: When a Magical Idea Comes Knocking, You Have Three Options', *Irish Times*, 7th January 2016, www.irishtimes.com/life-and-style/people/elizabeth-gilbert-when-a-magical-idea-comes-knocking-you-have-three-options-1.2474157.
43. Margaret J. Wheatley, *Finding Our Way: Leadership for an Uncertain Time* (San Francisco: Berrett-Koehler Publishers, 2007), 153. I have an American friend, a creativity coach, who deliberately takes her clients to this place of confusion and frustration in order to create breakthroughs.
44. 'Mixed Fluid Returns to its Original State', *New Scientist*, 25th August 2011, YouTube video, 1:31, www.youtube.com/watch?v=UpJ-kGII074.
45. Joseph Jaworski, *Synchronicity: The Inner Path of Leadership* (San Francisco: Berrett-Koehler Publishers, 2011), 148.

Chapter 11: Reality

1. William Blake, *The Marriage of Heaven and Hell* (Oxford: Benediction Classics, 2010), 17, Kindle.
2. Plato, 'The Allegory of the Cave', *Plato Six Pack*, trans. Benjamin Jowett (CreateSpace Independent Publishing Platform, 2017), 114, Kindle.

3. V.S. Ramachandran, *The Tell-Tale Brain: A Neuroscientist's Quest for What Makes Us Human* (Sydney: Cornerstone Digital, 2012), location 368, Kindle.
4. According to www.quora.com/Reality-is-merely-an-illusion-albeit-a-very-persistent-one-What-did-Albert-Einstein-mean-here (accessed 7th July 2022), this is a misquote. The actual quote, from a letter to the family of his lifelong friend Michele Besso, after learning of Besso's death in March 1955, was: 'People like us, who believe in physics, know the distinction between past, present, and future is only an illusion, albeit a stubborn one.' But I have chosen to stick with the quote as usually recounted.
5. As quoted in *The Observer* (25th January 1931), p.17, column 3.
6. As quoted in *The Observer* (11th January 1931); also in *Psychic Research* (1931), vol. 25, 91.
7. Anil Ananthaswamy, 'Do We Live in a Simulation? Chances Are About 50/50', *Scientific American*, 13th October 2020, www.scientificamerican.com/article/do-we-live-in-a-simulation-chances-are-about-50-50/.
8. Marc Abrahams, 'Experiments Show we Quickly Adjust to Seeing Everything Upside-down', *The Guardian*, 12th November 2012, www.theguardian.com/education/2012/nov/12/improbable-research-seeing-upside-down.
9. 'Erismann and Kohler: Inversion Goggles' uploaded by BioMotionLab, 11th November 2011, YouTube video, 12:41, www.youtube.com/watch?v=jKUVpBJalNQ.
10. Donald Hoffman, *The Case Against Reality: How Evolution Hid the Truth from Our Eyes* (London: Allen Lane, an imprint of Penguin, 2019), 163, Kindle.
11. Donald Hoffman, 'Is There an Infinite Mind?', 26th February 2020, in *The Waking Cosmos*, podcast, MP3 audio, https://soundcloud.com/wakingcosmos/is-there-an-infinite-mind-donald-hoffman-phd-the-waking-cosmos-podcast.
12. Iain McGilchrist, *The Divided Brain and the Search for Meaning.* (London: Yale University Press, 2012), location 254, Kindle.
13. David Brang and V. S. Ramachandran, 'Survival of the Synesthesia Gene: Why Do People Hear Colors and Taste Words?', *PLoS Biology* 9, no. 11 (November 2011): e1001205, https://doi.org/10.1371/journal.pbio.1001205.

14. Siri Carpenter, 'Everyday Fantasia: The World of Synesthesia', *American Psychological Association* 32, no. 3 (March 2001), www.apa.org/monitor/mar01/synesthesia.
15. See, for example, 'Memorandum for the Record, Problem: To Document Discussion Concerning Congresstional Inquiry Related To Parapsychology' (13th December 1977), Crest Stargate, CIA-RDP96-00788R001100030001-6 RIFPUB S 2, released 4th November, 2016. www.cia.gov/readingroom/docs/CIA-RDP96-00788R001100030001-6.pdf, accessed 8th July 2022.
16. Bernardo Kastrup, 'Mind Over Matter', in conversation with Jane Clark and Richard Gault, *Beshara Magazine*, issue 18, spring 2021, https://besharamagazine.org/science-technology/mind-over-matter/.
17. Bernardo Kastrup, *Why Materialism Is Baloney: How True Skeptics Know There Is No Death and Fathom Answers to Life, the Universe, and Everything*, 54, Kindle, published in 2014 by Iff Books, an imprint of John Hunt Publishing (www.johnhuntpublishing.com).
18. Ibid., 83, Kindle.
19. Donald Hoffman and Chetan Prakash, 'Objects of Consciousness', *Frontiers in Psychology* 5, no. 577 (17th June 2014), https://doi.org/10.3389/fpsyg.2014.00577.
20. Michael S. Gazzaniga, *Tales from Both Sides of the Brain: A Life in Neuroscience* (New York: Ecco Press, 2016).
21. David Wolman, 'The Split Brain: A Tale of Two Halves', *Nature* 483 (March 2012): 260–63, https://doi.org/10.1038/483260a.
22. Berit Brogaard, 'Split Brains', *Psychology Today*, 6th November 2012, www.psychologytoday.com/gb/blog/the-superhuman-mind/201211/split-brains.
23. Ed Yong, *I Contain Multitudes: The Microbes Within Us and a Grander View of Life* (London: Vintage, 2017), 3.
24. Rupert Sheldrake, *The Presence of the Past: Morphic Resonance and the Habits of Nature* (London: Icon Books, 2011), 20, Kindle.
25. Rupert Sheldrake, 'An Experimental Test of the Hypothesis of Formative Causation', *Rivista di Biologia – Biology Forum* 86, no. 3/4 (1992): 431–44, www.sheldrake.org/research/morphic-resonance/an-experimental-test-of-the-hypothesis-of-formative-causation.
26. Donald Hoffman, 'Is There an Infinite Mind?', 26th February 2020, in *The Waking Cosmos*, podcast, MP3 audio, https://soundcloud.com/wakingcosmos/is-there-an-infinite-mind-donald-hoffman-phd-the-waking-cosmos-podcast.

27. Scott Barry Kaufman, 'What Would Happen If Everyone Truly Believed Everything Is One?', *Scientific American*, 8th November 2018, https://blogs.scientificamerican.com/beautiful-minds/what-would-happen-if-everyone-truly-believed-everything-is-one/.
28. Kate J. Diebels and Mark R. Leary, 'The Psychological Implications of Believing that Everything is One', *The Journal of Positive Psychology* 14, no. 4 (2019), 463–73, https://doi.org/10.1080/17439760.2018.1484939.
29. Alex Evans, *The Myth Gap: What Happens When Evidence and Arguments Aren't Enough* (London: Eden Project Books, 2017).
30. Walter Sullivan, 'The Einstein Papers: A Man of Many Parts', *The New York Times*, 29th March 1972, www.nytimes.com/1972/03/29/archives/the-einstein-papers-a-man-of-many-parts-the-einstein-papers-man-of.html and https://timesmachine.nytimes.com/timesmachine/1972/03/29/issue.html.
31. *A Joseph Campbell Companion: Reflections on the Art of Living* (Joseph Campbell Foundation, 1991) 99–100.
32. Lachlan Brown, 'The Dalai Lama on Death', Hack Spirit, 30th April 2017, https://hackspirit.com/dalai-lama-explains-happens-die-can-prepared/.
33. For example: Bertrand Urien and William Kilbourne, 'On the Role of Materialism in the Relationship Between Death Anxiety and Quality of Life', *NA – Advances in Consumer Research* 35 (2008): 409–15, eds. Angela Y. Lee and Dilip Soman, Duluth, MN: Association for Consumer Research, www.acrwebsite.org/volumes/v35/naacr_vol35_0.pdf, and Andrew N. Christopher, *et al.*, 'Beliefs About One's Own Death, Personal Insecurity, and Materialism', *Personality and Individual Differences* 40, no. 3 (February 2006), 441–51, https://doi.org/10.1016/j.paid.2005.09.017.

Chapter 12: Ego

1. Alan Watts, *The Book: On the Taboo Against Knowing Who You Are* (London: Souvenir Press, 2009), 12.
2. Roz Savage, *The Gifts of Solitude: A Short Guide to Surviving and Thriving in Isolation* (UK, 2020), 134.

3. Jill Bolte Taylor, *My Stroke of Insight* (London: Hodder Paperbacks, 2009), 49.
4. Ibid., 61–7.
5. Ibid., 69.
6. Aldous Huxley, *The Doors of Perception* (first published in 1954 by Chatto & Windus), 6.
7. Michael Pollan, *How* to *Change Your Mind: The New Science of Psychedelics* (London: Allen Lane, 2018), 307.
8. Ibid., 308.
9. Eben Alexander, *Proof of Heaven: A Neurosurgeon's Journey into the Afterlife* (New York: Simon & Schuster, 2012), 72.
10. Michael Singer, *The Surrender Experiment: My Journey into Life's Perfection* (London: Yellow Kite, 2016), 5.
11. Erwin Schrödinger, *The Mystic Vision* as translated in *Quantum Questions: Mystical Writings of the World's Great Physicists*, ed. Ken Wilber (Boulder, CO: Shambhala, 2001)
12. Neale Donald Walsch, *Conversations with God, Book 3: An Uncommon Dialogue* (London: Hodder & Stoughton, 1999), 138.
13. Arthur Schopenhauer, *The World as Will and Representation*, translated by E.F.J. Payne, volumes 1 and 2 (New York: Dover Publications, 1969).
14. Will Storr, *Selfie: How the West Became Self-Obsessed* (London: Picador, 2018), 63.
15. Ibid., 70–71.
16. Albert Einstein, *The World As I See It*. (Digital Fire, 2019), 9, Kindle.
17. *Atomic Education Urged By Einstein, The New York Times*, 25th May 1946, https://timesmachine.nytimes.com/timesmachine/1946/05/25/100998236.html, accessed 8th July 2022.
18. Alan Watts., *The Book: On the Taboo Against Knowing Who You Are* (London: Souvenir Press, 2009), 130, Kindle.
19. Andrew Harvey, *Hidden Journey: A Spiritual Awakening* (London: Watkins Publishing, 2011).
20. Hunter S. Thompson, *The Proud Highway: Saga of a Desperate Southern Gentleman, 1955–1967* (London: Bloomsbury Paperbacks, 2011).
21. Neale Donald Walsch, *Conversations with God, Book 3: An Uncommon Dialogue* (London: Hodder & Stoughton, 1999), 186.
22. *Huai Nan Tzu* 18, 6a. A version of this story, as told by Lieh-Tzu, appears in Lin Yutang (1), 160 (as referenced by Alan Watts in *Tao: The Watercourse Way* (London: Souvenir Press, 2011).

23. Pema Chödrön, *When Things Fall Apart: Heart Advice for Difficult Times* (London: Element, 2005), 15.
24. Tristan Jones, *Outward Leg* (New York: Open Road Media, 2014), location 198, Kindle.
25. Benjamin Hoff, *The Tao of Pooh* (London: Egmont, 2012), 10.
26. Richard Rudd, *The Gene Keys: Embracing Your Higher Purpose (New Edition): Unlocking the Higher Purpose Hidden in Your DNA* (London: Watkins Publishing, 2015), 65.
27. C.G. Jung, *Visions: Notes of the Seminar Given in 1930–1934*, ed. Clare Douglas (Oxford: Routledge, 1998).
28. Roz Savage, 'Day 49: Oarally Challenged', Roz Savage, 18th January 2006, www.rozsavage.com/day-49-oarally-challenged/.
29. Hermann Hesse, *If the War Goes on: Reflections on War and Politics*, (New York: Farrar Straus & Giroux, 1971).
30. John Milton, *Paradise Lost, Book I* (Victoria, Australia: Leopold Classic Library, 2015), lines 221–70.
31. Richard Rudd, *The Gene Keys: Embracing Your Higher Purpose (New Edition): Unlocking the Higher Purpose Hidden in Your DNA* (London: Watkins Publishing, 2015), 68.
32. Roz Savage, *The Gifts of Solitude: A Short Guide to Surviving and Thriving in Isolation* (UK, 2020), 129.

Chapter 13: Preamble

1. Quoted by adrienne maree brown, *Emergent Strategy: Shaping Change, Changing Worlds* (Edinburgh: AK Press, 2017), 17, Kindle.
2. Margaret J. Wheatley, *Finding Our Way: Leadership for an Uncertain Time* (San Francisco: Berrett-Koehler Publishers, 2007), 153.
3. Milton Friedman, *Capitalism and Freedom* (London: University of Chicago Press, 2002), xiv.
4. David Wallace, 'A Large Solar Storm Could Knock Out the Power Grid and the Internet – An Electrical Engineer Explains How', *The Conversation*, 18th March 2022, https://theconversation.com/a-large-solar-storm-could-knock-out-the-power-grid-and-the-internet-an-electrical-engineer-explains-how-177982.

Chapter 14: Principles

1. Valerie Kaur, ‘What if This Darkness is not the Darkness of the Tomb, but the Darkness of the Womb?’, speech delivered at the Interfaith Watch Night Service at the Metropolitan AME Church, Washington State, 31st December 2016, https://speakola.com/ideas/tag/WATCH+NIGHT+SERVICE.
2. Alex Evans, *The Myth Gap: What Happens When Evidence and Arguments Aren't Enough* (London: Transworld Publishers, 2017), 104, Kindle.
3. See, for example, William Kremer and Claudia Hammond, *Abraham Maslow and the pyramid that beguiled business* (31st August 2013), BBC World Service, *BBC News Magazine*, www.bbc.co.uk/news/magazine-23902918.
4. Ilarion Merculieff, *et al.*, *Perspectives on Indigenous Issues: Essays on Science, Spirituality and the Power of Words* (Alaska: GCILL, 2018), quoted in ‘The Indigenous Art of Following Wisdom from the Heart’, Bioneers, accessed 9th July 2022, https://bioneers.org/the-indigenous-art-of-following-wisdom-from-the-heart-zeoz1903/.
5. Martin Luther King, Jr, ‘A Christmas Sermon on Peace’, *The Trumpet of Conscience* (Boston: Beacon Press, 2010), 70, 71.
6. Roz Savage, *The Gifts of Solitude: A Short Guide to Surviving and Thriving in Isolation* (UK, 2020), 129.
7. J. Shen-Miller, J.W. Schopf, G. Harbottle, R-J. Cao, S. Ouyang, K-S. Zhou, J.R. Southon, G-H .Liu, ‘Long-living lotus: Germination and soil-irradiation of centuries-old fruits, and cultivation, growth, and phenotypic abnormalities of offspring’, *American Journal of Botany*, 89 (2) (2002) 236–47, doi: 10.3732/ajb.89.2.236, accessed 13th July 2022, PMID 21669732.
8. Neale Donald Walsch, *Conversations with God, Book 3* (London: Hodder and Stoughton, 1999), 255, Kindle.
9. Don Beck, *Spiral Dynamics: Mastering Values, Leadership and Change* (Oxford: Wiley-Blackwell, 2005).

Chapter 15: Practicalities

1. Barbara Ransby, *Ella Baker & The Black Freedom Movement: A Radical Democratic Vision* (The University of North Carolina Press, 2003).
2. Robert Krulwich, 'How Human Beings Almost Vanished From Earth In 70,000 B.C.', *NPR*, 22nd October 2012, www.npr.org/sections/krulwich/2012/10/22/163397584/how-human-beings-almost-vanished-from-earth-in-70-000-b-c.
3. Rex Weyler, 'How Much of Earth's Biomass is Affected by Humans?', Greenpeace, 18th July 2018, www.greenpeace.org/international/story/17788/how-much-of-earths-biomass-is-affected-by-humans/.
4. 'How Chernobyl has Become an Unexpected Haven for Wildlife', UN Environment Programme, 16th September 2020, www.unep.org/news-and-stories/story/how-chernobyl-has-become-unexpected-haven-wildlife.
5. Robert J. Nicholls, *et al.*, 'Ranking of the World's Cities Most Exposed to Coastal Flooding Today and in the Future', OECD (2007), https://climate-adapt.eea.europa.eu/metadata/publications/ranking-of-the-worlds-cities-to-coastal-flooding/11240357, and Apeksha Bhateja, 'These 11 Cities Are Sinking. They Could Be Gone By 2100', Fodors, 17th September 2021, www.fodors.com/news/photos/these-11-cities-are-sinking-they-could-be-gone-by-2100.
6. With gratitude to Kim Stanley Robinson for the inspiration for the Adequacy concept, as he described during our conversation for the Sowing the Seeds of Change podcast.
7. 'Usury', *Britannica*, last modified 13th April 2018, www.britannica.com/topic/usury.
8. Robin Dunbar, *Grooming, gossip, and the Evolution of Language* (Cambridge, Massachusetts: Harvard University Press, 1988), 77.
9. Elinor Ostrom, *Governing the Commons: The Evolution of Institutions for Collective Action (Political Economy of Institutions and Decisions)* (Cambridge: Cambridge University Press, 1991).
10. Nick Bostrom, *Ethical Issues in Advanced Artificial Intelligence* (2003), www.researchgate.net/publication/229001428_Ethical_Issues_in_Advanced_Artificial_Intelligence, accessed 9th July 2022.

Chapter 16: Present

1. Roy T. Bennett, *The Light in the Heart: Inspirational Thoughts for Living Your Best Life* (Roy Bennett, 2021).
2. Paul Gilding, 'The One Degree War Plan', accessed 9th July 2022, www.paulgilding.com/one-degree-war-plan.
3. adrienne maree brown, *Emergent Strategy: Shaping Change, Changing Worlds* (Edinburgh: AK Press, 2017), 81, Kindle.
4. Irving Oyle, *The New American Medicine Show* (Santa Cruz, CA: Unity Press, 1979), 163.
5. From the Elders of the Hopi Nation at Oraibi, Arizona, 8th June 2000. As quoted by Margaret Wheatley, *Perseverance* (Oakland, CA: Berrett-Koehler Publishers, 2010).
6. Laura Tenenbaum, 'Digging In The Dirt Really Does Make People Happier', *Forbes*, 29th June 2020, www.forbes.com/sites/lauratenenbaum/2020/01/29/digging-in-the-dirt-really-does-make-people-happier/?sh=47cdd70831e1.
7. Buckminster Fuller, quoted in *Extreme Democracy*, eds. Mitch Ratcliffe and John Lebkowsky (2005), published under Creative Commons and available at http://extremedemocracy.com/ExtremeDemocracy.pdf.
8. Quoted by Richard Dawkins, *The God Delusion* (London: Bantam Press, 2006), 354.
9. 'Pascal's Wager', *Stanford Encyclopedia of Philosophy*, last modified 1st September 2017, https://plato.stanford.edu/entries/pascal-wager/.
10. 'Hopium', Urban Dictionary, last modified 25th September 2021, www.urbandictionary.com/define.php?term=Hopium.

Index

Also by Roz Savage

Rowing The Atlantic: Lessons Learned on the Open Ocean (2009)
Stop Drifting, Start Rowing: One Woman's Search for Happiness and Meaning Alone on the Pacific (2013)
The Gifts of Solitude: A Short Guide to Surviving and Thriving in Isolation (2020)